Fruit and Nuts

First Supplement to the Fifth Edition of

McCance and Widdowson's
The Composition of Foods

Fruit and Nuts

First Supplement to the Fifth Edition of

McCance and Widdowson's

The Composition of Foods

B. Holland, I. D. Unwin and D. H. Buss

The Royal Society of Chemistry
and
Ministry of Agriculture, Fisheries and Food

The Royal Society of Chemistry
Thomas Graham House
Science Park
Milton Road
Cambridge CB4 4WF
UK

Tel.:   (0223) 420066   Telex: 818293
Fax: (0223) 423429

ISBN 0-85186-386-8

Orders should be addressed to:
The Royal Society of Chemistry
Distribution Centre
Letchworth
Herts. SG6 1HN
UK

Tel.:   (0462) 672555   Telex: 825372
Fax: (0462) 480947

Xerox Ventura Publisher™ output photocomposed by
Goodfellow & Egan Phototypesetting Ltd, Cambridge

Printed in the United Kingdom by
The Bath Press, Lower Bristol Road, Bath

CONTENTS

ACKNOWLEDGEMENTS

A large number of people have helped at each stage in the preparation of this book.

Most of the new analyses of fruit, fruit products and nuts were performed at the Laboratory of the Government Chemist, Teddington. The analytical team was headed by Mrs G D Holcombe. Most of the non-starch polysaccharide fractions were determined at the Dunn Clinical Nutrition Centre by a team headed by Dr H N Englyst.

We are indebted to numerous manufacturers, retailers and other organisations for information on the range and composition of their products. In particular we would like to thank the Fresh Fruit and Vegetable Information Bureau, MPR Leedex Group Ltd, J Sainsbury plc and Safeway plc. We would also like to thank Mrs Azmina Govindji of the British Diabetic Association and members of the British Dietetic Association for their comments on useful inclusions to the tables.

The final preparation of this book was overseen by a committee which, besides the authors, comprised Miss P J Brereton (Northwick Park Hospital, Harrow), Dr M C Edwards (Campden Food and Drink Research Association, Chipping Campden), Dr A M Fehily (MRC Epidemiology Unit, Cardiff), Miss A A Paul (MRC Dunn Nutritional Laboratory, Cambridge), and Professor D A T Southgate (AFRC Institute of Food Research, Norwich).

We would also like to express our appreciation for all the help given to us by so many people in the Ministry of Agriculture, Fisheries and Food, the Royal Society of Chemistry and elsewhere who were involved in the work leading to the production of this book. In particular we would like to thank Mrs A A Welch (Royal Society of Chemistry) for her help in the latter stages of production of this supplement.

INTRODUCTION

This book contains detailed information on the nutritional value of a wide range of fruit, nuts and seeds. It complements and greatly extends the information given on these foods in the fifth edition of *The Composition of Foods*, which was published by the Royal Society of Chemistry in 1991.

The relationship between this supplement, the fifth edition and other books setting out the nutritional value of specific groups of foods in the UK is as follows. Between 1980 and 1991 the Ministry of Agriculture, Fisheries and Food (MAFF) and then MAFF together with the Royal Society of Chemistry produced five supplements to extend and update the nutrient information in the fourth edition of McCance and Widdowson's *The Composition of Foods* (Paul and Southgate, 1978). The first showed the amino acids and fatty acids in individual foods (Paul *et al.*, 1980), and subsequent supplements gave values for a wide range of nutrients in a large number of immigrant foods (Tan *et al.*, 1985), cereals and cereal products (Holland *et al.*, 1988), milk products and eggs (Holland *et al.*, 1989), and vegetables, herbs and spices (Holland *et al.*, 1991a). A selection of values from these supplements was included in the new fifth edition of *The Composition of Foods* (Holland *et al.*, 1991b), but because of the amount of detail in the supplements they and their associated computer files remain as essential reference works for those who need the additional information on the widest possible range of foods and nutrients.

The present supplement continues this tradition, and includes information on a much wider range of fruit and nuts and their products than could be covered in the fifth edition. The number of entries for fruit juices, 'exotic fruit' and canned fruit has been substantially increased, so that this book now shows the nutrients in 289 fruits and fruit products and 51 nuts and seeds compared with 120 and 27 respectively in the fifth edition. It also gives information on a number of nutrients that were not included in the fifth edition, including individual sugars, more components of fibre, sulfur, carotenoid fractions, vitamin E fractions, phytic acid and organic acids.

Methods   The selection of foods and of nutrient values has followed the general principles used in the preparation of previous supplements and in the fifth edition of *The Composition of Foods*. This supplement covers as many as possible of the fruit and nuts, and foods based upon them, that are currently used in the UK. Their nutrients have mainly been determined from new direct analyses although values from the scientific literature have been included where appropriate.

Literature values

Because many of the fruit and nuts eaten in the UK are grown and have been analysed in other countries, this supplement contains a significant number of values from the world's scientific literature. The values chosen give due recognition to the varieties and the production and storage methods used in the country of origin before they arrive in the UK. Where possible, the literature values were taken from studies that included full details of the samples; where suitable methods of analysis had been used; and where the results had been presented in sufficient detail. Nevertheless, there were many gaps, and for these fruit and nuts and for fruit that are normally grown in the UK a large number of new analyses were commissioned.

By analysis

Where little or nothing was known of the nutrients in an important raw or prepared product, or previous values related to varieties that are no longer widely available, arrangements were made for its direct analysis at the Laboratory of the Government Chemist. Additional analyses of non-starch polysaccharide fractions were undertaken at the Dunn Clinical Nutrition Centre in Cambridge. Detailed sampling and analytical protocols were devised for each item, and most of the foods were then bought from a wide variety of shops in London and Cambridge.

The analytical methods for the major nutrients were as described in the fifth edition of *The Composition of Foods*, while those for the nutrients not included in that book are given in the supplement on *Vegetables, Herbs and Spices* (Holland *et al.*, 1991a). Further details of each determination can be provided on request.

Arrangement of the tables

Food names and grouping

For ease of reference, the values for raw, cooked and canned fruit have been brought together, followed by a section for fruit juices, and finally one for nuts and seeds. Seeds used as spices were included in the supplement on *Vegetables, Herbs and Spices*. Although the foods have been listed alphabetically within each section, the first value for each is generally for the fresh or raw material, followed where appropriate by values for the cooked or processed versions of the food. 'Ready-to-eat' dried fruit are those which have been only partially dried. A number of fruit and nuts are known by different names in different cultures. They are listed under the most common name, but major alternatives have also been given and a wider variety of names is included in the combined index and food coding list at the end of this supplement. There is also a listing of taxonomic and alternative names to help with the identification of specific foods.

Numbering system

As in previous supplements, the foods have been numbered in sequence, with fruit and fruit products numbered from 1 to 289 and then, following a gap, nuts and seeds numbered from 801 to 851. Each food in the MAFF/RSC nutrient databank also has a unique two digit prefix to specify its food group. For this supplement, the prefix is '14'. The full code numbers for amla and walnuts weighed with shells, the first and last foods in this supplement, are thus 14-001 and 14-851, and these are the numbers that will be used in other databank applications. Because different numbers have been used for convenience in the fifth edition, a full list of codes is available from the Royal Society of Chemistry. This shows for all foods the numbers used in the fifth edition as well as those used in this and other supplements.

Description and main data sources

The information given under this heading indicates the number and nature of the samples taken for analysis. Some additional values for cooked fruit and related foods were calculated from these after correction for any differences in water content and nutrient losses, and, for some foods, the addition of sugar; and this is also indicated under this heading. The amounts of fruit, sugar and water are given both for mixtures that were analysed and for those where the nutrient values were derived by calculation. Where sugar was needed, 12g was in general used for each 100g fruit. Fruit juice concentrates have been characterised by their Brix value, the Brix scale being a commercial measure of specific gravity which reflects the sugar content of the juice.

For nutrient values derived from the literature, the major sources of information used are also shown under this heading.

Nutrients

The presentation of the nutrients is the same as that in the supplement on *Vegetables, Herbs and Spices*, to accommodate within four pages the nutritional information most appropriate to plant foods.

For most foods, the nutrient values refer to 100g of the edible part as described. However, many fruits and nuts are commonly weighed with their peel, stones or shell. For convenience an *additional* set of values has been given for the most widely used fruits and nuts showing the nutrients in 100g of these food as bought or served, weighed with the waste described, the consumable portion being shown in the edible proportion column. For the remaining foods this fraction of edible material is shown in the edible proportion column alongside the nutrient values for 100g of the edible portion.

Proximates: —The first page for each food begins with the proportion of edible matter in the food *as described*. For canned fruit, the juice or syrup has been assumed to be used and the edible proportion is given as 1.00. The weight of the fruit alone is shown in the supplementary table on proportions of fruit in cans before draining. The first page then continues with the amounts of water, total nitrogen, protein, fat, available carbohydrate expressed as its monosaccharide equivalent, and energy value of the food both in kilocalories and kilojoules, all expressed per 100g.

The protein content of fruit and fruit products was derived by multiplying the amount of nitrogen by 6.25, but different conversion factors were used for nuts and are shown in **Table 1**. Energy values were derived by multiplying the amounts

Table 1 *Factors used to convert the amount of nitrogen in nuts to protein[a]*

Almonds	5.18
Brazil nuts	5.41
Peanuts	5.41
All other nuts	5.30

[a] FAO/WHO (1973)

of protein, fat and carbohydrate by the factors in **Table 2**. No allowance has been made for the contribution made by citric acid, malic acid or other organic acids in fruit (the amounts of which are shown in a table, page116), although each gram of such acids could provide an additional 2.5 kcal or 10 kJ. A few of the carbohydrate values taken from the literature were not obtained by analysis but

Table 2 *Energy conversion factors*

	kcal/g	kJ/g
Protein	4	17
Fat	9	37
Available carbohydrate expressed as monosaccharide	3.75	16

estimated by difference, that is subtracting the other proximates from 100; these and the corresponding energy values have been presented as quoted, but in italics to distinguish them from the more accurate analytical values.

Carbohydrates and 'fibre': — The second page gives more details of the individual carbohydrates and fibre fractions. Some of the sucrose naturally present in fruit or added prior to canning or stewing will be hydrolysed to glucose and fructose, especially if the fruit is acidic; the extent of this hydrolysis has been taken as 10 per cent. The value for total sugars is the sum of the glucose, fructose, sucrose and maltose present and does not include any of the small amount of oligo-saccharides in some fruit and seeds. Wherever possible, however, the oligo-saccharides have been included in the total carbohydrate, and where this results in a value greater than the sum of the starch and the sugars this is indicated in a footnote.

As in previous editions, the amounts of sugars, starch and available carbohydrate are shown after conversion to their monosaccharide equivalents, but all fibre values are now the actual amounts of each component.

The relationships between the various forms and fractions of fibre are shown in **Table 3**.

Table 3 *Relationships between the dietary fibre fractions*

Cellulose

Insoluble non-cellulosic polysaccharides

} Insoluble fibre

Soluble non-cellulosic polysaccharides

} Soluble fibre

} Englyst fibre (non-starch polysaccharides)

} Southgate fibre[a] (unavailable carbohydrate)

'Lignin'

[a] The Southgate values are generally higher than NSP values because they include substances measuring as lignin and also because the enzymatic preparation used leaves some enzymatically resistant starch in the dietary fibre residue. A 'resistant starch' value can be obtained from the NSP procedures but because this uses different conditions and enzymes this may or may not be the same as the enzymatically resistant starch in the Southgate method

Minerals and vitamins: —The range of minerals and vitamins shown in the main tables is the same as in the fifth edition of *The Composition of Foods* except that values for sulfur have been added. Although most values were obtained by direct analysis, some of those for cooked and canned fruits were calculated from the corresponding raw foods. For these, any vitamin losses were estimated from the losses found on analysis of similar foods or from the losses quoted in the fifth edition which are shown in **Table 4**. The values for total carotene and for vitamin E have been corrected for the relative activities of the different fractions where these are known. Supplementary tables show further details of the individual carotenoids where data are available (page 108), and the amounts of the major tocopherols in a number of fruits and nuts (page 112).

Table 4 *Typical vitamin losses (%) on stewing fruit*

Thiamin	25
Riboflavin	25
Niacin	25
Vitamin B_6	20
Folate	80
Pantothenic acid	25
Biotin	25
Vitamin C	25

Appendices The total amounts of saturated, monounsaturated and polyunsaturated fatty acids in those foods containing 1 per cent of fat or more are given on page 106. And besides the supplementary tables showing the main carotene and vitamin E fractions, there are further tables showing phytic acid values (page 115), organic acids (page 116), the proportions of fruit in cans before draining (page 119) and alternative and taxonomic names (page 121).

Nutrient variability

It is important to appreciate that samples of the same or similar foods always vary somewhat in composition. Vitamin C is particularly variable, and the amount can differ between fruit from the same tree or bush and even between parts of the same fruit. Some nutrients tend to differ in a consistent way between varieties of a fruit or with season, as shown for apples. There will also be differences with the length and conditions of storage and the depth of peeling, and with cooking conditions, although any apparent differences between fruit grown by different methods (such as 'organic' methods) appear to be small and inconsistent, or due primarily to differences in water content. On the other hand, the nutrient composition of the fruit and juice in canned fruit are remarkably similar to each other (with the exception of carotenoids and fibre which are present in the fruit moiety in higher concentrations). It is not practical to give separate nutrient values for all these factors, so the tables show average values for most products. Where further values have been given, it remains possible that many of the differences are as much due to the above factors and to analytical variations as to real differences between the foods.

The amount of water and sugar added to stewed fruits depends on individual preference, and the amount of water lost by evaporation can also vary with the length of cooking. Fruits in this supplement were stewed with the minimum added water and, for calculated results, evaporative losses of 10% were applied.

For a number of dried fruits, the sum of the fat, protein, carbohydrate and water was substantially less than 100. This is most likely to be due to the presence of organic acids and skin components not measured during analysis.

A more comprehensive description of the factors to be taken into account in the proper use of food composition tables is given in the introduction to the fifth edition of *The Composition of Foods*. Users of the present supplement are advised to read them and take them to heart.

References to Introductory text

FAO/WHO (1973) *Energy and protein requirements*. Report of a Joint FAO/WHO *Ad Hoc* Expert Committee. FAO Nutrition Meetings Report Series, No. 52; WHO Technical Report Series, No. 522

Holland, B., Unwin, I. D., and Buss, D. H. (1988) *Third supplement to McCance and Widdowson's The Composition of Foods, 4th edition: Cereals and Cereal Products*, Royal Society of Chemistry, Cambridge

Holland, B., Unwin, I. D., and Buss, D. H. (1989) *Fourth supplement to McCance and Widdowson's The Composition of Foods, 4th edition: Milk Products and Eggs*, Royal Society of Chemistry, Cambridge

Holland, B., Unwin, I. D., and Buss, D. H. (1991a) *Fifth supplement to McCance and Widdowson's The Composition of Foods, 4th edition: Vegetables, Herbs and Spices*, Royal Society of Chemistry, Cambridge

Holland, B., Welch, A. A., Unwin, I. D., Buss, D. H., Paul, A. A. and Southgate, D. A. T. (1991b) *McCance and Widdowson's The Composition of Foods 5th edition*, Royal Society of Chemistry, Cambridge

Paul, A. A. and Southgate, D. A. T. (1978) *McCance and Widdowson's The Composition of Foods 4th edition*, HMSO, London

Paul, A. A., Southgate, D. A. T. and Russell, J. (1980) *First supplement to McCance and Widdowson's The Composition of Foods, 4th edition: Amino acid composition (mg per 100g food) and fatty acid composition (g per 100g food)*, HMSO, London

Tan, S. P., Wenlock, R. W., and Buss, D. H. (1985) *Second supplement to McCance and Widdowson's The Composition of Foods, 4th edition: Immigrant Foods*, HMSO, London

Symbols and abbreviations used in the tables

Symbols

0	None of the nutrient is present
Tr	Trace
N	The nutrient is present in significant quantities but there is no reliable information on the amount
( )	Estimated value
Italic text	Carbohydrate estimated 'by difference', and energy values based upon these quantities

Abbreviations

Gluc	Glucose
Fruct	Fructose
Sucr	Sucrose
Malt	Maltose
Lact	Lactose
Trypt	Tryptophan
Satd	Saturated
Monounsatd	Monounsaturated
Polyunsatd	Polyunsaturated
equiv	equivalents

Fruit

and

Fruit Juices

No. 14-	Food	Description and main data sources	Edible proportion	Water g	Total nitrogen g	Protein g	Fat g	Carbo-hydrate g	Energy value kcal	kJ
1	Amla	Ref. 4	0.89	81.8	0.08	0.5	0.1	13.7	58	243
2	Apples, cooking, raw, peeled	Bramley variety; flesh only	1.00	87.7	0.05	0.3	0.1	8.9	35	151
3	weighed with skin and core	Calculated from 2	0.73	63.1	0.04	0.2	0.1	6.4	26	109
4	stewed with sugar	Samples as raw. 1000g fruit, 100g water, 120g sugar	1.00	77.7	0.05	0.3	0.1	19.1	74	314
5	stewed without sugar	Samples as raw. 1000g fruit, 100g water and calculation from 4	1.00	87.5	0.04	0.3	0.1	8.1	33	138
6	baked with sugar, flesh and skin	Recipe, ref. 9. Calculation from 9; core removed	1.00	77.1	0.08	0.5	0.1	19.2	74	318
7	-, flesh only	Recipe. Ref. 9	1.00	76.5	0.08	0.5	0.1	20.1	78	332
8	-, weighed with skin	Recipe. Ref. 9	0.86	65.8	0.07	0.4	0.1	17.3	67	287
9	baked without sugar, flesh and skin	10 samples, cored and baked 180C, 30-40 mins	1.00	84.3	0.08	0.5	0.1	11.2	45	191
10	-, flesh only	Analysis and calculation from 9; cored before baking	1.00	85.0	0.08	0.5	0.1	10.7	43	183
11	-, weighed with skin	Calculated from 10	0.84	71.4	0.07	0.4	0.1	9.0	36	154
12	eating, average, raw	15 varieties; flesh and skin	1.00	84.5	0.06	0.4	0.1	11.8	47	199
13	average, raw, weighed with core	Calculated from 12	0.89	75.2	0.06	0.4	0.1	10.5	42	179
14	-, raw, peeled	Literature sources and calculation from 12; flesh only	1.00	85.4	0.06	0.4	0.1	11.2	45	190
15	-, raw, peeled, weighed with skin and core	Calculated from 14	0.76	64.9	0.05	0.3	0.1	8.5	34	145
16	dried	6 samples, 3 brands and calculation from 14	1.00	21.6	0.32	2.0	0.5	60.1	238	1014

14-001 to 14-016

Carbohydrate fractions, g per 100g

No. 14-	Food	Starch	Total sugars	Individual sugars					Dietary fibre		Fibre fractions			
				Gluc	Fruct	Sucr	Malt	Lact	Southgate method	Englyst method	Cellulose	Non-cellulosic polysaccharide		Lignin
												Soluble	Insoluble	
1	**Amla**	N	N	N	N	N	0	0	N	N	N	N	N	N
2	**Apples, cooking**, *raw, peeled*	Tr	8.9	2.0	5.9	1.0	0	0	2.2	1.6	0.6	0.6	0.4	Tr
3	*weighed with skin and core*	Tr	6.4	1.4	4.2	0.7	0	0	1.6	1.1	0.4	0.4	0.3	Tr
4	*stewed with sugar*	Tr	19.1	2.8	6.3	10.1	0	0	1.8	1.2	0.4	0.5	0.3	Tr
5	*stewed without sugar*	Tr	8.1	1.8	5.5	0.8	0	0	2.0	1.5	0.5	0.5	0.4	Tr
6	*baked with sugar, flesh and skin*	Tr	19.2	2.7	6.3	10.2	0	0	(2.6)	(1.8)	(0.7)	(0.7)	(0.5)	N
7	*-, flesh only*	Tr	20.1	2.5	5.9	11.6	0	0	2.0	1.7	0.6	0.6	0.4	Tr
8	*-, weighed with skin*	Tr	17.3	2.2	5.1	10.0	0	0	1.7	1.5	0.5	0.5	0.3	Tr
9	*baked without sugar, flesh and skin*	Tr	11.2	2.9	6.9	1.4	0	0	(2.8)	(2.0)	(0.8)	(0.8)	(0.5)	N
10	*-, flesh only*	Tr	10.7	2.8	6.6	1.3	0	0	2.2	1.9	0.7	0.7	0.5	Tr
11	*-, weighed with skin*	Tr	9.0	2.3	5.5	1.1	0	0	1.8	1.6	0.6	0.6	0.4	Tr
12	**eating**, *average, raw*	Tr	11.8[a]	1.7	6.2	3.9	0	0	(2.0)	1.8	0.6	0.7	0.4	0.1
13	*average, raw, weighed with core*	Tr	10.5	1.5	5.5	3.5	0	0	(1.8)	1.6	0.5	0.6	0.4	0.1
14	*-, raw, peeled*	Tr	11.2	1.6	5.9	3.7	0	0	1.8[b]	1.6	0.6	0.6	0.3	Tr
15	*-, raw, peeled, weighed with skin and core*	Tr	8.5	1.2	4.5	2.8	0	0	1.4	1.2	0.5	0.5	0.2	Tr
16	*dried*	Tr	60.1	8.6	31.7	19.9	0	0	9.7	9.7	2.8	4.7	2.2	N

[a] Levels ranged from 9.5 to 13.0g total sugars per 100g

[b] Apple peel contains 3.3g Southgate fibre per 100g

11

Fruit

No. 14-	Food	Na	K	Ca	Mg	P	Fe	Cu	Zn	S	Cl	Mn	Se	I
							mg						µg	
1	**Amla**	5	230	50	N	20	1.2	0.18	N	N	N	N	N	N
2	**Apples, cooking**, raw, peeled	2	88	4	3	7	0.1	0.02	Tr	3	2	Tr	Tr	Tr
3	weighed with skin and core	1	63	3	2	5	0.1	0.01	Tr	2	1	Tr	Tr	Tr
4	stewed with sugar	4	140	4	3	7	0.1	0.02	Tr	2	2	Tr	Tr	Tr
5	stewed without sugar	4	150	4	3	8	0.1	0.02	Tr	3	2	Tr	Tr	Tr
6	baked with sugar, flesh and skin	3	97	8	4	9	0.2	0.02	N	6	5	0.1	Tr	Tr
7	-, flesh only	3	93	8	5	10	0.2	0.02	N	6	6	0.2	Tr	Tr
8	-, weighed with skin	3	80	7	4	9	0.1	0.02	N	5	5	0.1	Tr	Tr
9	baked without sugar, flesh and skin	3	98	4	4	9	0.1	0.02	Tr	6	3	Tr	Tr	Tr
10	-, flesh only	3	94	4	4	9	0.1	0.02	Tr	6	3	Tr	Tr	Tr
11	-, weighed with skin	3	79	3	3	8	0.1	0.02	Tr	5	3	Tr	Tr	Tr
12	**eating**, average, raw	3	120	4	5	11	0.1	0.02	0.1	6	Tr	0.1	Tr	Tr
13	average, raw, weighed with core	3	110	4	4	10	0.1	0.02	0.1	5	Tr	0.1	Tr	Tr
14	-, raw, peeled	3	100	3	3	8	0.1	0.02	0.1	6	Tr	0.1	Tr	Tr
15	-, raw, peeled, weighed with skin and core	2	76	2	2	6	0.1	0.01	0.1	5	Tr	0.1	Tr	Tr
16	dried	16	540	16	16	43	0.5	0.11	0.5	N	1	0.5	Tr	Tr

Fruit

No. 14-	Food	Retinol µg	Carotene µg	Vitamin D µg	Vitamin E mg	Thiamin mg	Riboflavin mg	Niacin mg	Trypt 60 mg	Vitamin B_6 mg	Vitamin B_{12} µg	Folate µg	Panto-thenate mg	Biotin µg	Vitamin C mg
1	**Amla**	0	9	0	N	0.03	0.01	0.2	N	N	0	N	N	N	N
2	**Apples, cooking**, raw, peeled	0	(17)	0	0.27	0.04	0.02	0.1	0.1	0.06	0	5	Tr	1.2	14[a]
3	weighed with skin and core	0	(12)	0	0.19	0.03	0.01	0.1	0.1	0.04	0	4	Tr	0.9	10
4	stewed with sugar	0	(14)	0	0.22	0.01	0.01	0.1	0.1	0.05	0	Tr	Tr	0.8	10[b]
5	stewed without sugar	0	(15)	0	0.25	0.01	0.01	0.1	Tr	0.05	0	Tr	Tr	0.9	11
6	baked with sugar, flesh and skin	0	(16)	0	0.54	0.02	0.02	0.1	0.1	0.04	0	2	Tr	1.1	15
7	-, flesh only	0	(16)	0	0.24	0.02	0.02	0.1	0.1	0.04	0	2	Tr	1.1	12
8	-, weighed with skin	0	(14)	0	0.21	0.02	0.02	0.1	0.1	0.03	0	2	Tr	0.9	10
9	baked without sugar, flesh and skin	0	(18)	0	0.59	0.02	0.02	0.1	0.1	0.04	0	3	Tr	1.2	16
10	-, flesh only	0	(18)	0	0.27	0.02	0.02	0.1	0.1	0.04	0	3	Tr	1.2	13
11	-, weighed with skin	0	(15)	0	0.23	0.02	0.02	0.1	0.1	0.03	0	3	Tr	1.0	11
12	**eating**, average, raw	0	18	0	0.59	0.03	0.02	0.1	0.1	0.06	0	1	Tr	1.2	6[c]
13	average, raw, weighed with core	0	16	0	0.53	0.03	0.02	0.1	0.1	0.05	0	1	Tr	1.1	5
14	-, raw, peeled	0	17	0	0.27	0.03	0.02	0.1	0.1	0.06	0	1	Tr	1.1	4
15	-, raw, peeled, weighed with skin and core	0	13	0	0.21	0.02	0.01	0.1	0.1	0.05	0	1	Tr	0.8	3
16	dried	0	91	0	1.45	N	N	N	0.3	N	0	Tr	Tr	N	Tr

a Unpeeled cooking apples contain 20mg vitamin C per 100g

b Frozen apple slices, stewed with sugar contain 12mg vitamin C per 100g

c Levels ranged from 3 to 20mg vitamin C per 100g

No. 14-	Food	Description and main data sources	Edible proportion	Water g	Total nitrogen g	Protein g	Fat g	Carbo-hydrate g	Energy value kcal	kJ
17	**Apples, eating,** Cox's Pippin, raw	30 samples; flesh and skin	1.00	83.3	0.08	0.5	0.1	11.4	46	195
18	Cox's Pippin, *raw, weighed with core*	Calculated from 17	0.88	73.3	0.07	0.4	0.1	10.0	40	171
19	Golden Delicious, *raw*	30 samples; flesh and skin	1.00	85.5	0.05	0.3	0.2	10.8	43	185
20	Golden Delicious, *raw, weighed with core*	Calculated from 19	0.92	78.7	0.05	0.3	0.2	9.9	40	171
21	Granny Smith, *raw*	30 samples; flesh and skin	1.00	85.3	0.05	0.3	0.1	11.5	45	193
22	Granny Smith, *raw, weighed with core*	Calculated from 21	0.91	77.6	0.05	0.3	0.1	10.5	41	177
23	red dessert, *raw*	20 samples, assorted varieties; flesh and skin	1.00	83.8	0.05	0.3	0.1	13.0	51	217
24	red dessert, *raw, weighed with core*	Calculated from 23	0.90	75.4	0.05	0.3	0.1	11.7	46	196
25	**Apricots,** *raw*	18 samples; flesh and skin	1.00	87.2	0.14	0.9	0.1	7.2	31	134
26	*raw, weighed with stones*	Calculated from 25	0.92	80.2	0.13	0.8	0.1	6.6	29	123
27	*stewed with sugar*	1000g fruit, 100g water, 120g sugar; stones removed	1.00	76.9	0.11	0.7	0.1	18.3	72	308
28	*stewed with sugar, weighed with stones*	Calculated from 27	0.93	71.7	0.10	0.6	0.1	17.1	67	288
29	*stewed without sugar*	500g fruit, 75g water; stones removed	1.00	88.3	0.12	0.7	0.1	6.2	27	115
30	*stewed without sugar, weighed with stones*	Calculated from 29	0.93	82.1	0.11	0.7	0.1	5.8	25	108

Fruit continued

Carbohydrate fractions, g per 100g

No. 14-	Food	Starch	Total sugars	Individual sugars					Dietary fibre		Fibre fractions			
				Gluc	Fruct	Sucr	Malt	Lact	Southgate method	Englyst method	Cellulose	Non-cellulosic polysaccharide Soluble	Insoluble	Lignin
17	**Apples, eating,** Cox's Pippin, raw	Tr	11.4	1.3	5.6	4.5	0	0	(2.2)	2.0	0.7	0.9	0.4	0.1
18	Cox's Pippin, raw, weighed with core	Tr	10.0	1.1	4.9	4.0	0	0	(1.9)	1.8	0.6	0.8	0.3	0.1
19	Golden Delicious, raw	Tr	10.8	2.4	7.1	1.4	0	0	(1.9)	1.7	0.6	0.7	0.4	0.1
20	Golden Delicious, raw, weighed with core	Tr	9.9	2.2	6.5	1.3	0	0	(1.7)	1.6	0.5	0.6	0.4	0.1
21	Granny Smith, raw	Tr	11.5	2.6	6.0	2.9	0	0	(1.9)	1.7	0.6	0.7	0.4	0.1
22	Granny Smith, raw, weighed with core	Tr	10.5	2.4	5.5	2.6	0	0	(1.7)	1.5	0.5	0.6	0.4	0.1
23	red dessert, raw	Tr	13.0	3.3	7.8	1.9	0	0	(2.1)	1.9	0.6	0.9	0.4	0.1
24	red dessert, raw, weighed with core	Tr	11.7	3.0	7.0	1.7	0	0	(1.9)	1.7	0.5	0.8	0.4	0.1
25	**Apricots,** raw	0	7.2	1.6	0.9	4.6	0	0	1.9	1.7	0.5	1.0	0.2	0.1
26	raw, weighed with stones	0	6.6	1.5	0.8	4.2	0	0	1.7	1.6	0.5	0.9	0.2	0.1
27	stewed with sugar	0	18.3	2.2	2.1	14.0	0	0	1.8	1.6	0.5	0.9	0.2	0.1
28	stewed with sugar, weighed with stones	0	17.1	2.1	2.0	13.1	0	0	1.7	1.5	0.5	0.8	0.2	0.1
29	stewed without sugar	0	6.2	1.6	1.0	3.5	0	0	1.6	1.5	0.4	0.9	0.2	0.1
30	stewed without sugar, weighed with stones	0	5.8	1.5	0.9	3.3	0	0	1.5	1.4	0.4	0.8	0.2	0.1

Fruit *continued*

Inorganic constituents per 100g

No. 14-	Food	Na	K	Ca	Mg	P	Fe	Cu	Zn	S	Cl	Mn	Se	I
		mg											µg	
17	**Apples, eating,** Cox's Pippin, raw	3	130	4	6	12	0.2	Tr	Tr	6	Tr	Tr	Tr	Tr
18	Cox's Pippin, raw, weighed with core	3	110	3	5	11	0.2	Tr	Tr	5	Tr	Tr	Tr	Tr
19	Golden Delicious, raw	4	110	4	5	9	0.2	0.04	0.1	6	Tr	Tr	Tr	Tr
20	Golden Delicious, raw, weighed with core	4	100	4	5	8	0.2	0.04	0.1	5	Tr	0.1	Tr	Tr
21	Granny Smith, raw	2	120	4	4	9	0.1	0.02	Tr	6	1	0.1	Tr	Tr
22	Granny Smith, raw, weighed with core	1	110	4	4	8	0.1	0.02	Tr	5	1	Tr	Tr	Tr
23	red dessert, raw	2	110	4	5	10	0.1	0.04	Tr	6	1	0.1	Tr	Tr
24	red dessert, raw, weighed with core	1	99	4	5	8	0.1	0.04	Tr	5	Tr	Tr	Tr	Tr
25	**Apricots,** raw	1	270	15	11	20	0.5	0.06	0.1	6	3	Tr	(1)	N
26	raw, weighed with stones	2	250	14	10	18	0.5	0.05	0.1	5	3	0.1	(1)	N
27	stewed with sugar	1	190	12	7	7	0.8	0.05	0.1	4	2	Tr	(1)	N
28	stewed with sugar, weighed with stones	1	180	11	7	7	0.7	0.05	0.1	4	2	Tr	(1)	N
29	stewed without sugar	1	200	13	7	7	0.8	0.05	0.1	4	2	0.1	(1)	N
30	stewed without sugar, weighed with stones	1	190	12	7	7	0.7	0.05	0.1	4	2	0.1	(1)	N

No. 14-	Food	Retinol µg	Carotene µg	Vitamin D µg	Vitamin E mg	Thiamin mg	Ribo-flavin mg	Niacin mg	Trypt 60 mg	Vitamin B6 mg	Vitamin B12 µg	Folate µg	Panto-thenate mg	Biotin µg	Vitamin C mg
17	**Apples, eating,** Cox's Pippin, *raw*	0	(18)	0	0.59	0.03	0.03	0.2	0.1	0.08	0	4	Tr	1.2	9[a,b]
18	Cox's Pippin, *raw, weighed with core*	0	(16)	0	0.52	0.03	0.03	0.2	0.1	0.07	0	3	Tr	1.1	8
19	Golden Delicious, *raw*	0	15	0	0.59	0.03	0.03	0.1	0.1	0.11	0	1	Tr	1.2	4[b]
20	Golden Delicious, *raw, weighed with core*	0	13	0	0.54	0.03	0.03	0.1	0.1	0.10	0	1	Tr	1.1	4
21	Granny Smith, *raw*	0	5	0	0.59	0.04	0.02	0.1	0.1	0.08	0	1	Tr	1.2	4[b]
22	Granny Smith, *raw, weighed with core*	0	5	0	0.54	0.04	0.02	0.1	0.1	0.07	0	1	Tr	1.1	4
23	red dessert, *raw*	0	15	0	0.59	0.03	0.02	0.1	0.1	0.04	0	1	Tr	1.2	3[b]
24	red dessert, *raw, weighed with core*	0	13	0	0.53	0.03	0.02	0.1	0.1	0.04	0	1	Tr	1.1	3
25	**Apricots,** *raw*	0	405[c]	0	N	0.04	0.05	0.5	0.1	0.08	0	5	0.24	N	6
26	*raw, weighed with stones*	0	375	0	N	0.04	0.05	0.5	0.1	0.07	0	5	0.22	N	5
27	*stewed with sugar*	0	105	0	N	0.02	0.03	0.5	0.1	0.05	0	Tr	0.15	N	4
28	*stewed with sugar, weighed with stones*	0	98	0	N	0.02	0.03	0.5	0.1	0.05	0	Tr	0.14	N	4
29	*stewed without sugar*	0	110	0	N	0.03	0.03	0.3	0.1	0.05	0	1	0.15	N	4
30	*stewed without sugar, weighed with stones*	0	100	0	N	0.03	0.03	0.3	0.1	0.05	0	1	0.14	N	4

[a] Vitamin C levels in other varieties are Discovery 16mg, Sturmer 20mg, Worcester 5mg and Russet 8mg per 100g

[b] Storage may considerably affect vitamin C levels

[c] Levels ranged from 200 to 3370µg carotene per 100g

Fruit *continued*

14-031 to 14-046
Composition of food per 100g

No. 14-	Food	Description and main data sources	Edible proportion	Water (g)	Total nitrogen (g)	Protein (g)	Fat (g)	Carbo-hydrate (g)	Energy value (kcal)	Energy value (kJ)
31	**Apricots**, *dried*	No stones	1.00	14.7	0.77	4.8	0.7	43.4	188	802
32	*dried, stewed with sugar*	Calculated from 450g fruit, 770g water, 54g sugar	1.00	61.8	0.30	1.9	0.3	22.0	92	393
33	*-, stewed without sugar*	Calculated from 450g fruit, 770g water	1.00	65.0	0.32	2.0	0.3	17.8	77	328
34	*canned in syrup*	10 samples, 9 brands	1.00	80.0	0.07	0.4	0.1	16.1	63	268
35	*canned in juice*	10 samples, 5 brands	1.00	87.5	0.08	0.5	0.1	8.4	34	147
36	*ready-to-eat*	10 samples, no stones; semi-dried	1.00	29.7	0.63	4.0	0.6	36.5	158	674
37	**Avocado**, *average*	Average of Fuerte and Hass varieties	1.00	72.5[a]	0.30	1.9	19.5[b]	1.9[c]	190	784
38	*average, weighed with skin and stone*	Calculated from 37	0.71	51.5	0.21	1.3	13.8	1.3[c]	134	553
39	*Fuerte*	10 samples and literature sources; flesh only	1.00	73.1	0.34	2.1	19.3	1.9[c]	189	780
40	*weighed with skin and stone*	Calculated from 39	0.71	51.9	0.24	1.5	13.7	1.3[c]	134	553
41	*Hass*	7 samples and literature sources; flesh only	1.00	72.0	0.26	1.6	19.7	1.9[c]	191	787
42	*weighed with skin and stone*	Calculated from 41	0.71	51.1	0.18	1.1	14.0	1.3[c]	135	557
43	**Babaco**	Refs. 2, 12; flesh only	0.98	93.9	0.18	1.1	0.1	3.1	17	72
44	**Banana chips**	10 samples, 5 brands; crystallised	1.00	3.2	0.16	1.0	31.4	59.9	511	2137
45	**Bananas**	10 samples; flesh only	1.00	75.1	0.19	1.2	0.3	23.2	95	403
46	*weighed with skin*	Calculated from 45	0.66	49.6	0.13	0.8	0.2	15.3	62	266

[a] Water can range from 50 to 80g per 100g
[b] Fat can range from 10 to 40g per 100g
[c] Including mannoheptulose

Fruit *continued*

Carbohydrate fractions, g per 100g

No. 14-	Food	Starch	Total sugars	Individual sugars Gluc	Fruct	Sucr	Malt	Lact	Dietary fibre Southgate method	Englyst method	Fibre fractions Cellulose	Non-cellulosic polysaccharide Soluble	Insoluble	Lignin
31	**Apricots**, *dried*	0	43.4	20.8	10.0	12.6	0	0	21.6	7.7	2.4	4.6	0.7	0.7
32	*dried, stewed with sugar*	0	22.0	8.7	4.4	8.9	0	0	8.5	3.0	0.9	1.8	0.3	0.3
33	*-, stewed without sugar*	0	17.8	8.8	4.4	4.7	0	0	8.9	3.2	1.0	1.9	0.3	0.3
34	*canned in syrup*	0	16.1	6.7	5.8	3.7	0	0	1.2	0.9	0.3	0.4	0.2	Tr
35	*canned in juice*	0	8.4	3.0	4.1	1.4	0	0	(1.2)	0.9	0.3	0.4	0.2	Tr
36	*ready-to-eat*	0	36.5	17.5	8.4	10.6	0	0	18.1	6.3	2.0	3.8	0.5	0.5
37	**Avocado**, *average*	Tr	0.5[a]	0.3	0.1	0.1	0	0	N	3.4	1.1	1.6	0.7	0.2
38	*average, weighed with skin and stone*	Tr	0.4[a]	0.2	0.1	0.1	0	0	N	2.4	0.8	1.1	0.5	0.1
39	*Fuerte*	Tr	0.5[a]	0.3	0.1	0.1	0	0	N	3.4	1.1	1.6	0.7	0.2
40	*weighed with skin and stone*	Tr	0.4[a]	0.2	0.1	0.1	0	0	N	2.4	0.8	1.1	0.5	0.1
41	*Hass*	Tr	0.5[a]	0.3	0.2	Tr	0	0	N	3.4	1.1	1.6	0.7	0.2
42	*weighed with skin and stone*	Tr	0.4[a]	0.2	0.1	Tr	0	0	N	2.4	0.8	1.1	0.5	0.1
43	**Babaco**	0	3.1	1.3	1.1	0.7	0	0	N	N	N	N	N	N
44	**Banana chips**	37.8	22.1	0.1	0.3	21.7	0	0	4.8	1.7	0.5	1.1	0.1	0.3
45	**Bananas**	2.3[b]	20.9[b]	4.8	5.0	11.1	0	0	3.1[c]	1.1	0.3	0.7	0.1	0.2
46	*weighed with skin*	1.5	13.8	3.2	3.3	7.3	0	0	2.0	0.7	0.2	0.5	0.1	0.1

[a] Not including mannoheptulose

[b] These are proportions for yellow ripe bananas. The starch content falls and the sugar content rises on ripening

[c] Bananas contain significant amounts of resistant starch

No. 14-	Food	Na	K	Ca	Mg	P	Fe	Cu	Zn	S	Cl	Mn	Se	I
							mg						µg	
31	**Apricots**, *dried*	56	1880	92	65	120	4.1	0.40	0.7	160	35	0.4	(7)	N
32	*dried, stewed with sugar*	21	740	36	25	47	1.6	0.16	0.3	62	13	0.2	(3)	N
33	*-, stewed without sugar*	22	770	37	26	49	1.7	0.16	0.3	65	14	0.2	(3)	N
34	*canned in syrup*	10	150	19	5	8	0.2	Tr	0.1	1	2	Tr	Tr	7
35	*canned in juice*	5	170	21	7	12	0.4	0.03	0.1	(1)	2	Tr	Tr	7
36	*ready-to-eat*	14	1380	73	43	82	3.4	0.35	0.5	130	29	0.3	(5)	N
37	**Avocado**, *average*	6	450	11	25	39	0.4	0.19	0.4	19	6	0.2	Tr	2
38	*average, weighed with skin and stone*	4	320	8	18	28	0.3	0.13	0.3	13	4	0.1	Tr	1
39	*Fuerte*	9	430	11	25	43	0.4	0.35	0.6	19	9	0.2	Tr	2
40	*weighed with skin and stone*	6	310	8	18	31	0.3	0.25	0.4	13	6	0.1	Tr	1
41	*Hass*	3	470	11	24	35	0.4	Tr	0.2	(19)	3	(0.1)	Tr	(2)
42	*weighed with skin and stone*	2	330	8	17	25	0.3	Tr	0.1	(13)	2	(0.1)	Tr	(1)
43	**Babaco**	2	140	11	6	14	0.4	N	0.1	N	N	N	N	N
44	**Banana chips**	5	470	13	66	61	0.8	0.10	0.4	N	N	0.6	N	N
45	**Bananas**	1	400	6	34	28	0.3	0.10	0.2	13	79	0.4	(1)	8
46	*weighed with skin*	1	270	4	22	18	0.2	0.66	0.1	9	52	0.3	(1)	5

No. 14-	Food	Retinol μg	Carotene μg	Vitamin D μg	Vitamin E mg	Thiamin mg	Ribo-flavin mg	Niacin mg	Trypt 60 mg	Vitamin B6 mg	Vitamin B12 μg	Folate μg	Panto-thenate mg	Biotin μg	Vitamin C mg
31	**Apricots**, *dried*	0	645	0	N	Tr	0.20	3.0	0.6	0.17	0	14	0.70	N	Tr
32	*dried, stewed with sugar*	0	255	0	N	Tr	0.06	0.9	0.2	0.05	0	1	0.21	N	Tr
33	*-, stewed without sugar*	0	265	0	N	Tr	0.06	0.9	0.3	0.06	0	1	0.22	N	Tr
34	*canned in syrup*	0	155	0	N	0.01	0.01	0.3	0.1	(0.06)	0	2	(0.06)	(0.4)	5
35	*canned in juice*	0	210	0	N	0.02	0.01	0.3	0.1	0.06	0	2	0.06	0.4	14
36	*ready-to-eat*	0	545	0	N	Tr	0.16	2.3	0.5	0.14	0	11	0.58	N	1
37	**Avocado**, *average*	0	16	0	3.20	0.10	0.18	1.1	0.3	0.36	0	11	1.10	3.6	6
38	*average, weighed with skin and stone*	0	11	0	2.27	0.07	0.13	0.8	0.2	0.26	0	8	0.78	2.6	4
39	*Fuerte*	0	16	0	3.20	0.12	0.19	1.0	0.4	0.34	0	13	1.10	3.6	5
40	*weighed with skin and stone*	0	11	0	2.27	0.09	0.13	0.7	0.3	0.24	0	9	0.78	2.6	3
41	*Hass*	0	16	0	3.20	0.08	0.17	1.2	0.3	0.38	0	8	1.10	3.6	7
42	*weighed with skin and stone*	0	11	0	2.27	0.06	0.12	0.9	0.2	0.27	0	6	0.78	2.6	5
43	**Babaco**	0	175	0	N	0.03	0.04	0.7	N	N	0	N	N	N	26
44	**Banana chips**	0	N	0	N	(0.04)	(0.07)	(0.8)	0.2	0.32	0	(15)	(0.39)	(2.8)	Tr
45	**Bananas**	0	21	0	0.27	0.04	0.06	0.7	0.2	0.29	0	14	0.36	2.6	11
46	*weighed with skin*	0	14	0	0.18	0.03	0.04	0.5	0.1	0.19	0	9	0.24	1.7	7

Fruit *continued*

Composition of food per 100g

No. 14-	Food	Description and main data sources	Edible proportion	Water g	Total nitrogen g	Protein g	Fat g	Carbo-hydrate g	Energy value kcal	kJ
47	**Bilberries**	Literature sources, whole fruit	0.98	85.9	0.10	0.6	0.2	6.9	30	128
48	**Blackberries,** *raw*	Cultivated and wild berries; whole fruit	1.00	85.0	0.14	0.9	0.2	5.1	25	104
49	*stewed with sugar*	Calculated from 700g fruit, 210g water, 84g sugar	1.00	78.9	0.11	0.7	0.2	13.8	56	239
50	*stewed without sugar*	Calculated from 700g fruit, 210g water	1.00	87.2	0.12	0.8	0.2	4.4	21	88
51	**Blackberry and apple,** *stewed with sugar*	Calculated from 400g blackberries, 550g apples, 175g water, 115g sugar	1.00	77.1	0.08	0.5	0.1	18.1	70	300
52	*stewed without sugar*	Calculated from 400g blackberries, 550g apples, 175g water	1.00	87.6	0.09	0.6	0.1	6.4	27	116
53	**Blackcurrants,** *raw*	Whole fruit, stalks removed	0.98	77.4	0.15	0.9	Tr	6.6	28	121
54	*stewed with sugar*	Calculated from 700g fruit, 210g water, 84g sugar	1.00	72.9	0.12	0.7	Tr	15.0	58	252
55	*stewed without sugar*	Calculated from 700g fruit, 210g water	1.00	80.7	0.13	0.8	Tr	5.6	24	103
56	**canned in juice**	4 samples, 2 brands	1.00	84.0	0.13	0.8	Tr	7.6	31	135
57	**canned in syrup**	3 samples of the same brand (Hartley's)	1.00	75.0	0.11	0.7	Tr	18.4	72	306
58	**Boysenberries,** canned in syrup	Ref. 3	1.00	76.3	0.16	1.0	0.1	20.4	88	368
59	**Carambola**	Analysis and literature sources; ends trimmed	0.97	91.4	0.08	0.5	0.3	7.3	32	136
60	**Cashew fruit**	Literature sources; flesh only	0.82	86.0	0.14	0.9	0.4	6.8	33	139
61	**Cherries,** *raw*	10 samples of black and red cherries; flesh and skin	1.00	82.8	0.14	0.9	0.1	11.5	48	203
62	*raw, weighed with stones*	Calculated from 61	0.83	68.7	0.12	0.7	0.1	9.5	39	168

No. Food 14-	Starch	Total sugars	Individual sugars					Dietary fibre		Cellulose	Non-cellulosic polysaccharide		Lignin
			Gluc	Fruct	Sucr	Malt	Lact	Southgate method	Englyst method		Soluble	Insoluble	
47 **Bilberries**	0	6.9	3.3	3.3	0.4	0	0	(2.5)	1.8	0.7	0.5	0.6	1.3
48 **Blackberries**, raw	0	5.1	2.5	2.6	Tr	0	0	6.6	3.1	1.2	1.0	0.9	N
49 stewed with sugar	0	13.8	2.5	2.5	8.9	0	0	5.2	2.4	0.9	0.8	0.7	N
50 stewed without sugar	0	4.4	2.1	2.2	Tr	0	0	5.6	2.6	1.0	0.9	0.8	N
51 **Blackberry and apple**, stewed with sugar	Tr	18.0	2.5	4.2	11.4	0	0	3.6	1.9	0.7	0.7	0.5	N
52 stewed without sugar	Tr	6.4	2.1	3.9	0.5	0	0	4.0	2.1	0.8	0.7	0.6	N
53 **Blackcurrants**, raw	0	6.6	3.0	3.4	0.3	0	0	7.8	3.6	0.8	1.6	1.2	1.8
54 stewed with sugar	0	15.0	2.8	3.2	9.0	0	0	6.1	2.8	0.6	1.3	0.9	1.4
55 stewed without sugar	0	5.6	2.5	2.9	0.2	0	0	6.7	3.1	0.7	1.4	1.0	1.5
56 canned in juice	0	7.6	3.4	4.0	0.2	0	0	(4.2)	3.1	0.6	1.1	1.4	(1.0)
57 canned in syrup	0	18.4	8.1	8.3	2.0	0	0	(3.6)	2.6	0.5	0.9	1.2	(0.9)
58 **Boysenberries**, canned in syrup	0	20.4	N	N	N	0	0	N	1.6	0.8	0.4	0.4	0.3
59 **Carambola**	0.2	7.1	3.1	3.2	0.8	0	0	1.7	1.3	0.3	0.6	0.3	0.2
60 **Cashew fruit**	Tr	6.8	N	N	N	0	0	N	N	N	N	N	N
61 **Cherries**, raw	0	11.5	5.9	5.3	0.2	0	0	1.5	0.9	0.2	0.5	0.2	Tr
62 raw, weighed with stones	0	9.5	4.9	4.4	0.2	0	0	1.2	0.7	0.1	0.4	0.2	Tr

Fruit continued

Inorganic constituents per 100g

No. 14-	Food	Na	K	Ca	Mg	P	Fe	Cu	Zn	S	Cl	Mn	Se	I
							mg						µg	
47	**Bilberries**	3	88	12	5	14	0.5	0.08	0.2	13	5	0.3	Tr	N
48	**Blackberries**, raw	2	160	41	23	31	0.7	0.11	0.2	9	22	1.4	Tr	N
49	stewed with sugar	1	130	32	17	24	0.5	0.09	0.2	7	17	1.1	Tr	N
50	stewed without sugar	1	140	35	19	26	0.6	0.09	0.2	7	18	1.2	Tr	N
51	**Blackberry and apple**, stewed with sugar	1	100	18	10	15	0.3	0.06	0.1	4	9	0.6	Tr	N
52	stewed without sugar	1	110	20	11	17	0.4	0.06	0.1	5	10	0.6	Tr	N
53	**Blackcurrants**, raw	3	370	60	17	43	1.3	0.14	0.3	33	15	0.3	N	N
54	stewed with sugar	2	290	47	13	33	1.0	0.11	0.3	25	11	0.2	N	N
55	stewed without sugar	2	320	51	14	36	1.1	0.12	0.3	28	12	0.3	N	N
56	**canned in juice**	Tr	190	26	13	29	5.2	0.04	0.1	N	20	0.2	N	N
57	**canned in syrup**	Tr	130	25	10	27	4.7	0.02	0.1	N	17	0.2	N	N
58	**Boysenberries**, canned in syrup	3	90	18	11	10	0.4	0.07	0.2	N	N	0.3	N	N
59	**Carambola**	2	150	5	6	15	0.6	0.12	0.1	N	N	0.1	N	N
60	**Cashew fruit**	7	120	9	N	25	0.8	N	N	N	N	N	N	N
61	**Cherries**, raw	1	210	13	10	21	0.2	0.07	0.1	7	Tr	0.1	(1)	Tr
62	raw, weighed with stones	1	170	11	8	17	0.2	0.06	0.1	6	Tr	0.1	(1)	Tr

Fruit *continued*

No. 14-	Food	Retinol µg	Carotene µg	Vitamin D µg	Vitamin E mg	Thiamin mg	Ribo-flavin mg	Niacin mg	Trypt 60 mg	Vitamin B6 mg	Vitamin B12 µg	Folate µg	Panto-thenate mg	Biotin µg	Vitamin C mg
47	**Bilberries**	0	30	0	N	0.03	0.03	0.4	0.1	0.05	0	6	0.25	1.1	17
48	**Blackberries**, *raw*	0	80	0	2.37	0.02	0.05	0.5	0.1	0.05	0	34	0.25	0.4	15
49	*stewed with sugar*	0	62	0	1.85	0.01	0.03	0.3	0.1	0.03	0	5	0.15	0.2	9
50	*stewed without sugar*	0	68	0	2.03	0.01	0.03	0.3	0.1	0.03	0	5	0.16	0.3	10
51	**Blackberry and apple**, *stewed with sugar*	0	(40)	0	1.08	0.02	0.02	0.2	0.1	0.04	0	3	0.08	0.5	9
52	*stewed without sugar*	0	(44)	0	1.20	0.02	0.02	0.2	0.1	0.04	0	3	0.09	0.5	10
53	**Blackcurrants**, *raw*	0	100	0	1.00	0.03	0.06	0.3	0.1	0.08	0	N	0.40	2.4	200[a]
54	*stewed with sugar*	0	78	0	0.78	0.02	0.04	0.2	0.1	0.05	0	N	0.23	1.4	115
55	*stewed without sugar*	0	85	0	0.83	0.02	0.04	0.2	0.1	0.05	0	N	0.26	1.5	130
56	**canned in juice**	0	(29)	0	0.54	0.01	0.03	0.2	0.1	0.11	0	4	(0.14)	(0.9)	37
57	**canned in syrup**	0	(29)	0	(0.54)	(0.01)	(0.03)	(0.2)	0.1	(0.11)	0	(4)	(0.14)	(0.9)	57
58	**Boysenberries**, canned in syrup	0	24	0	N	0.03	0.03	0.2	0.2	0.04	0	34	0.13	N	6
59	**Carambola**	0	37	0	N	0.03	0.03	0.4	N	N	0	N	N	N	31
60	**Cashew fruit**	0	115	0	N	0.03	0.03	0.4	N	N	0	N	N	N	220
61	**Cherries**, *raw*	0	25	0	0.13	0.03	0.03	0.2	0.1	0.05	0	5	0.26	0.4	11
62	*raw, weighed with stones*	0	21	0	0.11	0.02	0.02	0.2	0.1	0.04	0	4	0.22	0.3	9

[a] Levels ranged from 150 to 230mg vitamin C per 100g

25

Fruit *continued*

Composition of food per 100g

No. 14-	Food	Description and main data sources	Edible proportion	Water g	Total nitrogen g	Protein g	Fat g	Carbohydrate g	Energy value kcal	kJ
63	**Cherries,** *stewed with sugar*	Calculated from 900g fruit, 200g water, 108g sugar	1.00	75.1	0.11	0.7	0.1	21.0	82	350
64	*stewed with sugar, weighed with stones*	Calculated from 63	0.86	64.6	0.09	0.6	0.1	18.1	71	303
65	*stewed without sugar*	Calculated from 900g fruit, 200g water	1.00	84.9	0.12	0.8	0.1	10.1	42	177
66	*stewed without sugar, weighed with stones*	Calculated from 65	0.85	72.2	0.10	0.6	0.1	8.6	36	152
67	*canned in syrup*	10 samples, red and black	1.00	77.8	0.09	0.5	Tr	18.5	71	305
68	**glacé**	10 samples, 8 brands; red and multicoloured	1.00	23.6	0.07	0.4	Tr	66.4	251	1069
69	**West Indian**	Literature sources; flesh only	0.75	90.5	0.06	0.4	0.4	5.0	24	102
70	**Cherry pie filling**	10 samples, 7 brands	1.00	75.8	0.07	0.4	Tr	21.5	82	412
71	**Clementines**	10 samples; flesh only	1.00	87.5	0.14	0.9	0.1	8.7	37	158
72	*weighed with peel and pips*	Calculated from 71	0.75	65.6	0.11	0.7	0.1	6.5	28	120
73	**Cranberries**	Analysis and literature sources; whole fruit	1.00	87.0	0.06	0.4	0.1	3.4	15	65
74	**Currants**	10 samples, 9 brands	1.00	15.7	0.37	2.3	0.4	67.8	267	1139
75	**Custard apple/Bullock's heart**	Literature sources; flesh only	0.56	75.8	0.24	1.5	0.4	(16.1)	(70)	(298)
76	**Custard apple/Sugar apple**	Literature sources; flesh only	0.51	73.2	0.26	1.6	0.3	(16.1)	(69)	(296)

No. 14-	Food	Starch	Total sugars	Individual sugars Gluc	Fruct	Sucr	Malt	Lact	Dietary fibre Southgate method	Englyst method	Fibre fractions Cellulose	Non-cellulosic polysaccharide Soluble	Insoluble	Lignin
63	**Cherries**, *stewed with sugar*	0	21.0	5.2	4.8	10.9	0	0	1.2	0.7	0.2	0.4	0.2	Tr
64	*stewed with sugar, weighed with stones*	0	18.1	4.5	4.1	9.4	0	0	1.0	0.6	0.2	0.3	0.2	Tr
65	*stewed without sugar*	0	10.1	5.2	4.7	0.2	0	0	1.3	0.8	0.2	0.4	0.2	Tr
66	*stewed without sugar, weighed with stones*	0	8.6	4.4	4.0	0.1	0	0	1.1	0.7	0.2	0.3	0.2	Tr
67	*canned in syrup*	0	18.5	7.3	6.6	4.3	0.3	0	(0.7)	0.6	0.2	0.3	0.1	Tr
68	**glacé**	0	66.4	23.6	12.7	9.5	20.7	0	(1.5)	0.9	0.2	0.5	0.2	Tr
69	**West Indian**	0	5.0	N	N	N	0	0	N	N	N	N	N	N
70	**Cherry pie filling**	3.9	17.6	7.2	6.4	3.9	0.1	0	N	0.4	0.1	N	N	Tr
71	**Clementines**	0	8.7	1.5	1.7	5.6	0	0	(1.7)	1.2	0.3	0.7	0.2	Tr
72	*weighed with peel and pips*	0	6.5	1.1	1.3	4.2	0	0	(1.3)	0.9	0.2	0.5	0.1	Tr
73	**Cranberries**	0	3.4	2.2	1.2	Tr	0	0	3.8	3.0	1.3	1.1	0.6	N
74	**Currants**	0	67.8	34.4	33.3	Tr	0	0	5.9	1.9	0.7	1.0	0.2	N
75	**Custard apple/Bullock's heart**	(1.2)	(14.9)	(5.4)	(5.6)	(3.9)	0	0	N	N	N	N	N	N
76	**Custard apple/Sugar apple**	(1.2)	(14.9)	(5.4)	(5.6)	(3.9)	0	0	N	N	N	N	N	N

Fruit *continued*

Inorganic constituents per 100g

No. 14-	Food	Na	K	Ca	Mg	P	Fe (mg)	Cu	Zn	S	Cl	Mn	Se (µg)	I
63	**Cherries**, *stewed with sugar*	Tr	170	10	7	16	0.2	0.06	0.1	5	Tr	0.1	1	Tr
64	*stewed with sugar, weighed with stones*	Tr	140	9	6	14	0.1	0.05	0.1	4	Tr	0.1	1	Tr
65	*stewed without sugar*	Tr	180	11	8	18	0.2	0.06	0.1	6	Tr	0.1	1	Tr
66	*stewed without sugar, weighed with stones*	Tr	160	9	7	15	0.1	0.05	0.1	5	Tr	0.1	1	Tr
67	*canned in syrup*	8	120	15	7	13	2.9	Tr	Tr	N	N	0.1	(1)	N[a]
68	**glacé**	27	24	56	5	9	0.9	0.08	0.1	21	N	Tr	Tr	N[a]
69	**West Indian**	5	110	11	18	15	0.2	N	N	N	10	N	N	N
70	**Cherry pie filling**	30	75	28	5	17	2.6	Tr	Tr	N	N	0.1	N	N
71	**Clementines**	4	130	31	10	18	0.1	0.01	0.1	(10)	(2)	Tr	N	N
72	*weighed with peel and pips*	3	97	23	7	13	0.1	0.01	0.1	(7)	(1)	Tr	N	N
73	**Cranberries**	2	95	12	7	11	0.7	0.05	0.2	11	Tr	0.1	Tr	N
74	**Currants**	14	720	93	30	71	1.3	0.81	0.3	31	16	0.7	N	N
75	**Custard apple/Bullock's heart**	5	440	24	21	25	0.6	0.15	(0.2)	N	(37)	N	N	N
76	**Custard apple/Sugar apple**	7	290	21	34	35	0.6	(0.15)	(0.2)	N	37	N	N	N

[a] Iodine from erythrosine is present but largely unavailable

Fruit *continued*

No. 14-	Food	Retinol µg	Carotene µg	Vitamin D µg	Vitamin E mg	Thiamin mg	Ribo-flavin mg	Niacin mg	Trypt 60 mg	Vitamin B6 mg	Vitamin B12 µg	Folate µg	Panto-thenate mg	Biotin µg	Vitamin C mg
63	**Cherries**, *stewed with sugar*	0	19	0	0.10	0.02	0.02	0.1	0.1	0.03	0	Tr	0.15	0.2	7
64	*stewed with sugar, weighed with stones*	0	16	0	0.09	0.02	0.02	0.1	0.1	0.03	0	Tr	0.13	0.2	6
65	*stewed without sugar*	0	21	0	0.11	0.02	0.02	0.1	0.1	0.04	0	Tr	0.17	0.3	7
66	*stewed without sugar, weighed with stones*	0	18	0	0.09	0.02	0.02	0.1	0.1	0.03	0	Tr	0.14	0.3	6
67	*canned in syrup*	0	17	0	(0.06)	0.02	0.01	0.1	Tr	(0.22)	0	5	(0.08)	(0.1)	1
68	**glacé**	0	7	0	Tr	Tr	Tr	Tr	Tr	Tr	0	Tr	Tr	Tr	Tr
69	**West Indian**	0	(315)	0	N	0.02	0.07	0.4	Tr	0.01	0	N	0.31	N	1720[a]
70	**Cherry pie filling**	0	18	0	N	0.02	0.01	0.2	0.1	N	0	2	N	N	1
71	**Clementines**	0	75	0	N	0.09	0.04	0.3	0.1	(0.07)	0	33	(0.20)	N	54
72	*weighed with peel and pips*	0	57	0	N	0.07	0.03	0.2	0.1	(0.05)	0	25	(0.15)	N	41
73	**Cranberries**	0	22	0	N	0.03	0.02	0.1	0.1	0.07	0	2	0.22	N	13
74	**Currants**	0	6	0	N	0.16	0.05	0.9	0.2	0.23	0	4	0.07	4.8	Tr
75	**Custard apple/Bullock's heart**	0	29	0	N	0.09	0.08	0.5	0.1	0.22	0	N	0.13	N	21
76	**Custard apple/Sugar apple**	0	4	0	N	0.11	0.11	0.8	0.1	0.20	0	N	0.23	N	36

[a] Levels ranged from 1020 to 2560mg vitamin C per 100g

Fruit continued

Composition of food per 100g

No. 14-	Food	Description and main data sources	Edible proportion	Water g	Total nitrogen g	Protein g	Fat g	Carbo-hydrate g	Energy value kcal	kJ
77	**Damsons**, *raw*	Flesh and skin	1.00	77.5	0.08	0.5	Tr	9.6	38	162
78	*raw, weighed with stones*	Calculated from 77	0.90	69.7	0.07	0.5	Tr	8.6	34	146
79	*stewed with sugar*	Calculated from 1050g fruit, 210g water, 126g sugar	1.00	70.6	0.07	0.4	Tr	19.3	74	316
80	*stewed with sugar, weighed with stones*	Calculated from 79	0.92	64.9	0.06	0.4	Tr	17.6	68	288
81	*stewed without sugar*	Calculated from 1050g fruit, 210g water	1.00	79.5	0.07	0.5	Tr	8.7	34	147
82	*stewed without sugar, weighed with stones*	Calculated from 81	0.91	72.3	0.06	0.4	Tr	7.9	31	133
83	**Dates**, *raw*	5 samples; flesh and skin	1.00	60.7	0.24	1.5	0.1	31.3	124	530
84	*raw, weighed with stones*	Calculated from 83	0.86	52.2	0.21	1.3	0.1	26.9	107	456
85	*dried*	Flesh and skin	1.00	14.6	0.53	3.3	0.2	68.0	270	1151
86	*dried, weighed with stones*	Calculated from 85	0.84	12.3	0.45	2.8	0.2	57.1	227	969
87	**Dried mixed fruit**	Calculated as sultanas 49%, currants 24%, raisins 18% and peel 9%	1.00	15.5	0.37	2.3	0.4	68.1	268	1144
88	**Durian**	Literature sources; flesh only	0.22	62.9	0.39	2.4	1.4	28.4	136	575
89	**Elderberries**	Refs. 3, 7; whole fruit	1.00	80.3	0.11	0.7	0.5	7.4	35	149
90	**Feijoa**	Ref. 8; flesh only	0.51	85.0	0.15	0.9	0.5	10.3	47	199

No.	Food	Starch	Total sugars	Individual sugars					Dietary fibre		Fibre fractions			
				Gluc	Fruct	Sucr	Malt	Lact	Southgate method	Englyst method	Cellulose	Non-cellulosic polysaccharide		Lignin
14-												Soluble	Insoluble	
77	**Damsons**, *raw*	0	9.6	5.2	3.4	1.0	0	0	3.7	(1.8)	(0.3)	(1.2)	(0.3)	(0.2)
78	*raw, weighed with stones*	0	8.6	4.7	3.1	0.9	0	0	3.3	(1.6)	(0.3)	(1.1)	(0.3)	(0.2)
79	*stewed with sugar*	0	19.3	4.9	3.4	11.1	0	0	3.0	(1.5)	(0.2)	(1.0)	(0.2)	(0.2)
80	*stewed with sugar, weighed with stones*	0	17.6	4.4	3.1	10.2	0	0	2.8	(1.4)	(0.2)	(0.9)	(0.2)	(0.1)
81	*stewed without sugar*	0	8.7	4.8	3.1	0.8	0	0	3.4	(1.6)	(0.3)	(1.1)	(0.3)	(0.2)
82	*stewed without sugar, weighed with stones*	0	7.9	4.3	2.9	0.7	0	0	3.1	(1.5)	(0.3)	(1.0)	(0.3)	(0.2)
83	**Dates**, *raw*	0	31.3	16.2	15.1	Tr	0	0	3.6	1.8	0.6	0.5	0.7	0.9
84	*raw, weighed with stones*	0	26.9	13.9	13.0	Tr	0	0	3.1	1.5	0.5	0.4	0.6	0.8
85	*dried*	0	68.0	(35.4)	(32.6)	Tr	0	0	7.8	4.0	1.3	1.2	1.5	2.0
86	*dried, weighed with stones*	0	57.1	(29.7)	(27.4)	Tr	0	0	6.5	3.4	1.1	1.0	1.3	1.7
87	**Dried mixed fruit**	0	68.1	33.3	31.6	0.8	2.3	0	5.6	2.2	0.8	1.1	0.3	N
88	**Durian**	5.2	23.2	N	N	N	0	0	4.0	N	N	N	N	0.9
89	**Elderberries**	0	7.4	N	N	0.3	0	0	N	N	N	N	N	N
90	**Feijoa**	0.3	10.0	N	N	N	0	0	N	N	N	N	N	Tr

No. 14-	Food	Na	K	Ca	Mg	P	Fe	Cu	Zn	S	Cl	Mn	Se	I
							mg						µg	
77	**Damsons**, *raw*	2	290	24	11	16	0.4	0.08	(0.1)	6	Tr	N	Tr	N
78	*raw, weighed with stones*	2	260	22	10	14	0.4	0.07	(0.1)	5	Tr	N	Tr	N
79	*stewed with sugar*	1	240	19	9	13	0.3	0.07	(0.1)	4	Tr	N	Tr	N
80	*stewed with sugar, weighed with stones*	1	220	17	8	12	0.3	0.06	(0.1)	4	Tr	N	Tr	N
81	*stewed without sugar*	1	260	21	9	14	0.4	0.07	(0.1)	5	Tr	N	Tr	N
82	*stewed without sugar, weighed with stones*	1	240	19	8	13	0.3	0.06	(0.1)	5	Tr	N	Tr	N
83	**Dates**, *raw*	7	410	24	24	28	0.3	0.12	0.2	23	210	0.2	Tr	N
84	*raw, weighed with stones*	6	350	21	21	24	0.3	0.10	0.2	20	180	0.2	(1)	N
85	*dried*	10	700	45	41	60	1.3	0.26	0.4	51	370	0.3	(1)	N
86	*dried, weighed with stones*	8	590	38	34	50	1.1	0.22	0.3	43	310	0.3	(3)	N
87	**Dried mixed fruit**	48	880	73	29	73	2.2	0.47	0.4	33	13	0.4	(3)	N
88	**Durian**	1	600	14	33	52	0.9	N	N	N	N	N	N	N
89	**Elderberries**	1	290	37	N	48	1.6	N	N	N	N	N	N	N
90	**Feijoa**	3	130	5	7	13	0.1	0.02	0.1	12	21	Tr	N	N

No. 14-	Food	Retinol µg	Carotene µg	Vitamin D µg	Vitamin E mg	Thiamin mg	Ribo-flavin mg	Niacin mg	Trypt 60 mg	Vitamin B6 mg	Vitamin B12 µg	Folate µg	Panto-thenate mg	Biotin µg	Vitamin C mg
77	**Damsons**, *raw*	0	(295)	0	0.70	0.10	0.03	0.3	0.1	(0.05)	0	(3)	0.27	0.1	(5)
78	*raw, weighed with stones*	0	(265)	0	0.60	0.09	0.03	0.3	0.1	(0.05)	0	(3)	0.24	0.1	(5)
79	*stewed with sugar*	0	(240)	0	0.57	0.06	0.02	0.2	0.1	(0.03)	0	Tr	0.17	0.1	(3)
80	*stewed with sugar, weighed with stones*	0	(220)	0	0.52	0.05	0.02	0.2	0.1	(0.03)	0	Tr	0.16	0.1	(3)
81	*stewed without sugar*	0	(270)	0	0.64	0.07	0.02	0.2	0.1	(0.04)	0	Tr	0.18	0.1	(3)
82	*stewed without sugar, weighed with stones*	0	(245)	0	0.58	0.06	0.02	0.2	0.1	(0.04)	0	Tr	0.16	0.1	(3)
83	**Dates**, *raw*	0	(18)	0	N	0.06	0.07	0.7	0.7	0.12	0	25	0.21	N	14
84	*raw, weighed with stones*	0	(15)	0	N	0.05	0.06	0.6	0.6	0.10	0	21	0.18	N	12
85	*dried*	0	(40)	0	N	0.07	0.09	1.8	1.5	0.19	0	13	0.78	N	Tr
86	*dried, weighed with stones*	0	(34)	0	N	0.06	0.08	1.5	1.3	0.16	0	11	0.65	N	Tr
87	**Dried mixed fruit**	0	9	0	N	0.10	0.05	0.7	0.2	0.22	0	15	0.09	3.9	Tr
88	**Durian**	0	11	0	N	0.29	0.29	1.1	N	N	0	N	N	N	41
89	**Elderberries**	0	(360)	0	N	0.07	0.07	1.0	0.1	0.24	0	17	0.16	1.8	27
90	**Feijoa**	0	31	0	0.18	Tr	0.01	0.2	N	0.05	0	N	N	N	29

14-091 to 14-108

Composition of food per 100g

No. 14-	Food	Description and main data sources	Edible proportion	Water g	Total nitrogen g	Protein g	Fat g	Carbo-hydrate g	Energy value kcal	Energy value kJ
91	**Figs**, raw	Analysis and literature sources; whole green fruit	0.98	84.6	0.21	1.3	0.3	9.5	43	185
92	dried	Analysis and literature sources; whole fruit	1.00	16.8	0.57	3.6	1.6	52.9	227	967
93	dried, stewed with sugar	Calculated from 450g fruit, 450g water, 54g sugar	1.00	50.1	0.30	1.9	0.8	34.3	143	612
94	-, stewed without sugar	Calculated from 450g fruit, 450g water	1.00	53.8	0.32	2.0	0.9	29.4	126	537
95	ready-to-eat	6 samples; semi-dried	1.00	23.6	0.52	3.3	1.5	48.6	209	889
96	**Fruit cocktail**, canned in juice	10 samples, 6 brands	1.00	86.9	0.07	0.4	Tr	7.2	29	122
97	canned in syrup	Analysis and calculation from dissection proportions[a]	1.00	81.8	0.06	0.4	Tr	14.8	57	244
98	**Fruit pie filling**	10 samples, 7 brands. Assorted flavours	1.00	79.5	0.06	0.4	Tr	20.1	77	328
99	**Fruit salad**, homemade	Calculated from equal proportions of bananas, oranges, apples, pears and grapes	1.00	82.3	0.11	0.7	0.1	13.8	55	237
100	**Gooseberries**, cooking, raw	Tops and tails removed	0.91	90.1	0.18	1.1	0.4	3.0	19	81
101	stewed with sugar	1000g fruit, 150g water, 120g sugar	1.00	82.1	0.11	0.7	0.3	12.9	54	229
102	stewed without sugar	500g fruit, 100g water and calculation from 101	1.00	90.6	0.15	0.9	0.3	2.5	16	66
103	dessert, raw	Tops and tails removed	0.93	87.5	0.11	0.7	0.3	9.2	40	170
104	canned in syrup	4 samples, 2 brands	1.00	78.9	0.06	0.4	0.2	18.5	73	310
105	**Grapefruit**, raw	10 samples; flesh only	1.00	89.0	0.13	0.8	0.1	6.8	30	126
106	raw, weighed with peel and pips	Calculated from 105	0.68	60.5	0.09	0.5	0.1	4.6	20	86
107	canned in juice	10 samples, 8 brands	1.00	88.6	0.09	0.6	Tr	7.3	30	120
108	canned in syrup	10 samples	1.00	81.8	0.08	0.5	Tr	15.5	60	257

[a] Proportions as pears 42%, peaches 41%, pineapple 8%, grapes 5% and cherries 4%

Fruit *continued*

Carbohydrate fractions, g per 100g

No. 14-	Food	Starch	Total sugars	Gluc	Fruct	Sucr	Malt	Lact	Dietary fibre Southgate method	Dietary fibre Englyst method	Cellulose	Non-cellulosic polysaccharide Soluble	Non-cellulosic polysaccharide Insoluble	Lignin
91	**Figs**, *raw*	0	9.5	5.2	4.1	0.3	0	0	2.3	1.5	0.3	0.9	0.3	N
92	*dried*	0	52.9	28.6	22.7	1.6	0	0	12.4	7.5	2.4	4.0	1.1	N
93	*dried, stewed with sugar*	0	34.3	15.4	12.3	6.7	0	0	6.5	3.9	1.3	2.1	0.6	N
94	*-, stewed without sugar*	0	29.4	15.9	12.7	0.8	0	0	6.9	4.2	1.3	2.2	0.6	N
95	*ready-to-eat*	0	48.6	26.2	20.8	1.5	0.3	0	11.4	6.9	2.2	3.7	1.0	N
96	**Fruit cocktail**, *canned in juice*	0	7.2	3.2	3.5	0.5	0	0	1.0	1.0	0.4	0.4	0.2	Tr
97	*canned in syrup*	0	14.8	6.1	6.4	1.9	0	0	1.0	1.0	0.4	0.4	0.2	Tr
98	**Fruit pie filling**	5.5	14.6	5.2	5.5	3.9	0	0	1.6	1.0	0.3	0.5	0.2	Tr
99	**Fruit salad**, *homemade*	0.5	13.3	3.7	5.7	3.9	0	0	(1.5)	1.5	0.5	0.7	0.3	0.1
100	**Gooseberries**, *cooking, raw*	0	3.0	1.3	1.6	Tr	0	0	2.9	2.4	0.7	0.9	0.9	N
101	*stewed with sugar*	0	12.9	2.4	2.6	7.8	0	0	2.3	1.9	0.5	0.7	0.7	N
102	*stewed without sugar*	0	2.5	1.1	1.3	Tr	0	0	2.4	2.0	0.6	0.7	0.7	N
103	*dessert, raw*	0	9.2	4.3	4.5	0.4	0	0	3.1	2.4	0.7	0.9	0.9	N
104	*canned in syrup*	0	18.5	7.5	7.3	3.7	0	0	1.7	1.7	0.5	0.8	0.4	N
105	**Grapefruit**, *raw*	0	6.8	2.1	2.3	2.4	0	0	(1.6)	1.3	0.3	0.9	0.1	0.1
106	*raw, weighed with peel and pips*	0	4.6	1.4	1.6	1.6	0	0	(1.1)	0.9	0.2	0.6	0.1	0.1
107	*canned in juice*	0	7.3	3.6	3.4	0.3	0	0	(0.8)	0.4	0.1	0.3	Tr	0.1
108	*canned in syrup*	0	15.5	6.7	6.9	1.9	0	0	(0.9)	0.6	0.1	0.4	0.1	0.1

Fruit *continued*

Inorganic constituents per 100g

No. 14-	Food	Na	K	Ca	Mg	P	Fe	Cu	Zn	S	Cl	Mn	Se	I
							mg						µg	
91	**Figs**, *raw*	3	200	38	15	15	0.3	0.06	0.3	13	18	0.1	Tr	N
92	*dried*	62	970	250	80	89	4.2	0.30	0.7	81	170	0.5	Tr	N
93	*dried, stewed with sugar*	32	510	130	41	46	2.2	0.16	0.4	42	89	0.3	Tr	N
94	*-, stewed without sugar*	34	540	140	44	49	2.3	0.17	0.4	44	94	0.3	Tr	N
95	*ready-to-eat*	57	890	230	73	82	3.9	0.27	0.6	74	160	0.5	Tr	N
96	**Fruit cocktail**, canned in juice	3	95	9	7	14	0.4	0.04	0.1	1	2	0.1	Tr	N[a]
97	canned in syrup	3	95	5	5	9	0.3	0.02	0.1	1	3	0.1	Tr	N[a]
98	**Fruit pie filling**	43	84	30	5	15	1.0	0.02	Tr	N	45	0.1	Tr	N
99	**Fruit salad**, *homemade*	2	210	16	12	18	0.2	0.07	0.1	8	16	0.1	1	2
100	**Gooseberries**, cooking, *raw*	2	210	28	7	34	0.3	0.06	0.1	16	7	0.1	Tr	Tr
101	*stewed with sugar*	7	140	19	6	22	0.3	0.07	0.1	13	5	0.3	Tr	Tr
102	*stewed without sugar*	2	170	23	6	28	0.3	0.05	0.1	13	6	0.1	Tr	Tr
103	*dessert, raw*	1	170	19	9	19	0.6	0.06	0.1	14	11	0.1	Tr	Tr
104	*canned in syrup*	2	66	12	4	12	0.2	Tr	0.3	8	3	0.1	Tr	Tr
105	**Grapefruit**, *raw*	3	200	23	9	20	0.1	0.02	Tr	7	3	Tr	(1)	N
106	*raw, weighed with peel and pips*	2	140	16	6	14	0.1	0.01	Tr	5	2	Tr	(1)	N
107	*canned in juice*	10	72	22	8	16	0.3	0.01	Tr	N	(5)	Tr	Tr	N
108	*canned in syrup*	10	79	17	7	13	0.7	(0.01)	0.4	N	(5)	Tr	Tr	N

a Iodine from erythrosine is present but largely unavailable

No. 14-	Food	Retinol μg	Carotene μg	Vitamin D μg	Vitamin E mg	Thiamin mg	Ribo-flavin mg	Niacin mg	Trypt 60 mg	Vitamin B6 mg	Vitamin B12 μg	Folate μg	Panto-thenate mg	Biotin μg	Vitamin C mg
91	**Figs**, *raw*	0	(150)	0	N	0.03	0.03	0.4	0.2	0.08	0	N	0.22	N	2
92	*dried*	0	(64)	0	N	0.08	0.10	0.8	0.5	0.26	0	9	0.51	N	1
93	*dried, stewed with sugar*	0	(33)	0	N	0.03	0.04	0.3	0.3	0.11	0	Tr	0.20	N	Tr
94	*-, stewed without sugar*	0	(35)	0	N	0.03	0.04	0.3	0.3	0.12	0	Tr	0.21	N	Tr
95	*ready-to-eat*	0	(59)	0	N	0.07	0.09	0.7	0.4	0.24	0	8	0.47	N	1
96	**Fruit cocktail**, canned in juice	0	54	0	N	0.01	0.01	0.3	0.1	0.04	0	6	0.05	0.3	14
97	canned in syrup	0	(54)	0	N	0.02	0.01	0.4	0.1	0.03	0	5	0.05	0.1	4
98	**Fruit pie filling**	0	17	0	N	0.01	0.01	0.2	0.1	N	0	3	N	N	7
99	**Fruit salad**, *homemade*	0	20	0	0.32	0.05	0.03	0.3	0.1	0.11	0	9	0.17	1.1	16
100	**Gooseberries**, cooking, raw	0	110	0	0.37	0.03	0.03	0.3	0.2	0.02	0	(8)	0.29	0.5	14
101	stewed with sugar	0	41	0	0.29	0.01	0.02	0.2	0.1	0.01	0	6	0.17	0.3	11
102	stewed without sugar	0	43	0	0.31	0.01	0.02	0.2	0.1	0.01	0	6	0.18	0.3	11
103	dessert, raw	0	62	0	0.37	0.03	0.03	0.3	0.1	0.02	0	(8)	0.29	0.1	26
104	canned in syrup	0	(18)	0	(0.20)	(0.01)	(0.01)	(0.1)	0.1	(0.01)	0	(2)	(0.01)	Tr	27
105	**Grapefruit**, *raw*	0	17[a]	0	(0.19)	0.05	0.02	0.3	0.1	0.03	0	26	0.28	(1.0)	36
106	*raw, weighed with peel and pips*	0	11	0	(0.13)	0.03	0.01	0.2	0.1	0.02	0	18	0.19	(0.7)	24
107	*canned in juice*	0	Tr	0	(0.10)	0.04	0.01	0.3	0.1	(0.02)	0	6	(0.12)	(1.0)	33
108	*canned in syrup*	0	Tr	0	(0.11)	0.04	0.01	0.2	0.1	0.02	0	4	0.12	1.0	30

[a] Pink varieties contain approximately 280μg carotene per 100g

Fruit *continued*

No. 14-	Food	Description and main data sources	Edible proportion	Water g	Total nitrogen g	Protein g	Fat g	Carbo-hydrate g	Energy value kcal	kJ
109	**Grapes**, *average*[a]	10 samples; white, black and seedless	1.00	81.8	0.06	0.4	0.1	15.4	60	257
110	*weighed with pips*	Calculated from 109	0.95	77.7	0.06	0.4	0.1	14.6	57	244
111	**Greengages**, *raw*	Flesh and skin	1.00	82.0	0.12	0.8	0.1	9.7	41	173
112	*raw, weighed with stones*	Calculated from 111	0.95	77.9	0.11	0.7	0.1	9.2	38	163
113	*stewed with sugar*	Calculated from 1000g fruit, 100g water, 120g sugar	1.00	72.4	0.11	0.7	0.1	20.7	81	347
114	*stewed with sugar, weighed with stones*	Calculated from 113	0.96	69.5	0.11	0.7	0.1	19.9	78	334
115	*stewed without sugar*	1000g fruit, 100g water; stones removed	1.00	83.3	0.11	0.7	0.1	8.7	36	155
116	*stewed without sugar, weighed with stones*	Calculated from 115	0.95	79.1	0.10	0.6	0.1	8.3	34	147
117	**Grenadillas**	Analysis and literature sources; flesh and seeds	0.51	74.7	0.45	2.8	0.3	7.5	42	179
118	**Guava**, *raw*	Literature sources	1.00	84.7	0.13	0.8	0.5	5.0	26	112
119	*raw, weighed with skin and pips*	Calculated from 118	0.90	76.2	0.12	0.7	0.5	4.5	24	102
120	*canned in syrup*	10 samples	1.00	77.6	0.06	0.4	Tr	15.7	60	258
121	**Jambu fruit**	Refs. 4, 11; flesh only	0.83	85.3	0.11	0.7	0.3	12.4	55	234
122	**Jujube**	Literature sources; flesh only	0.92	76.7	0.23	1.5	0.3	19.7	87	371
123	**Kiwi fruit**	Analysis and literature sources, flesh and seeds	1.00	84.0	0.18	1.1	0.5	10.6	49	207
124	*weighed with skin*	Calculated from 123	0.86	72.2	0.15	1.0	0.4	9.1	42	177

[a] Few significant differences reported between varieties

Fruit continued

Carbohydrate fractions, g per 100g

No. 14-	Food	Starch	Total sugars	Individual sugars					Dietary fibre		Fibre fractions			
				Gluc	Fruct	Sucr	Malt	Lact	Southgate method	Englyst method	Cellulose	Non-cellulosic polysaccharide Soluble	Non-cellulosic polysaccharide Insoluble	Lignin
109	**Grapes**, average	0	15.4	7.6	7.8	0.1	0	0	0.8	0.7	0.3	0.4	Tr	0.1
110	weighed with pips	0	14.6	7.2	7.4	0.1	0	0	0.8	0.7	0.3	0.4	Tr	0.1
111	**Greengages**, raw	0	9.7	4.7	2.2	2.7	0	0	2.3	2.1	0.3	1.3	0.5	(0.2)
112	raw, weighed with stones	0	9.2	4.5	2.1	2.6	0	0	2.2	2.0	0.3	1.2	0.5	(0.2)
113	stewed with sugar	0	20.7	4.9	2.7	13.0	0	0	2.1	1.9	0.3	1.2	0.5	(0.2)
114	stewed with sugar, weighed with stones	0	19.9	4.8	2.6	12.4	0	0	2.0	1.8	0.3	1.1	0.5	(0.2)
115	stewed without sugar	0	8.7	4.3	2.1	2.2	0	0	2.1	1.9	0.3	1.2	0.5	(0.2)
116	stewed without sugar, weighed with stones	0	8.3	4.1	2.0	2.1	0	0	2.0	1.8	0.3	1.1	0.5	(0.2)
117	**Grenadillas**	0	7.5	2.9	2.2	2.4	0	0	14.3	3.3	1.0	1.1	1.2	N
118	**Guava**, raw	0.1	4.9	2.1	2.3	0.5	0	0	4.7	3.7	1.5	1.0	1.2	1.3
119	raw, weighed with skin and pips	0.1	4.4	1.9	2.1	0.5	0	0	4.2	3.3	1.3	0.9	1.1	1.2
120	canned in syrup	Tr	15.7	5.5	6.5	3.7	0	0	3.2	3.0	1.4	0.6	1.0	N
121	**Jambu fruit**	Tr	N	N	N	N	0	0	N	N	N	N	N	N
122	**Jujube**	Tr	*19.7*	N	N	N	0	0	N	N	N	N	N	N
123	**Kiwi fruit**	0.3	10.3	4.6	4.3	1.3	0	0	N	1.9	0.7	0.9	0.3	0.2
124	weighed with skin	0.3	8.9	4.0	3.7	1.1	0	0	N	1.6	0.6	0.8	0.3	0.2

Fruit *continued*

No. 14-	Food	Na	K	Ca	Mg	P	Fe	Cu	Zn	S	Cl	Mn	Se	I
							mg						µg	
109	**Grapes**, *average*	2	210	13	7	18	0.3	0.12	0.1	8	Tr	0.1	(1)	1
110	*weighed with pips*	2	200	12	7	17	0.3	0.11	0.1	8	Tr	0.1	(1)	1
111	**Greengages**, *raw*	1	310	17	8	23	0.4	0.08	0.1	3	1	N	Tr	N
112	*raw, weighed with stones*	1	290	16	8	22	0.4	0.08	0.1	3	1	N	Tr	N
113	*stewed with sugar*	Tr	280	15	7	20	0.4	0.07	0.1	2	Tr	N	Tr	N
114	*stewed with sugar, weighed with stones*	Tr	270	14	7	19	0.3	0.07	0.1	2	Tr	N	Tr	N
115	*stewed without sugar*	1	280	15	7	21	0.4	0.07	0.1	3	1	N	Tr	N
116	*stewed without sugar, weighed with stones*	1	270	14	7	20	0.4	0.07	0.1	3	1	N	Tr	N
117	**Grenadillas**	28	350	16	39	54	1.1	0.12	(0.8)	19	37	N	N	N
118	**Guava**, *raw*	5	230	13	12	25	0.4	0.10	0.2	14	4	0.1	N	N
119	*raw, weighed with skin and pips*	5	210	12	11	23	0.4	0.09	0.2	13	4	0.1	N	N
120	*canned in syrup*	7	120	8	6	11	0.5	0.10	0.4	N	10	N	N	N
121	**Jambu fruit**	26	55	11	35	12	1.1	0.23	N	13	8	N	N	N
122	**Jujube**	4	240	30	10	22	0.6	0.07	0.1	N	N	0.1	N	N
123	**Kiwi fruit**	4	290	25	15	32	0.4	0.13	0.1	16	39	0.1	N	N
124	*weighed with skin*	3	250	21	13	27	0.3	0.11	0.1	14	33	0.1	N	N

Fruit continued

No. 14-	Food	Retinol µg	Carotene µg	Vitamin D µg	Vitamin E mg	Thiamin mg	Ribo-flavin mg	Niacin mg	Trypt 60 mg	Vitamin B6 mg	Vitamin B12 µg	Folate µg	Panto-thenate mg	Biotin µg	Vitamin C mg
109	**Grapes**, average	0	17	0	Tr	0.05	0.01	0.2	Tr	0.10	0	2	0.05	0.3	3
110	weighed with pips	0	16	0	Tr	0.05	0.01	0.2	Tr	0.09	0	2	0.05	0.3	3
111	**Greengages**, raw	0	95	0	(0.70)	0.05	0.04	0.6	0.1	(0.05)	0	(3)	(0.20)	Tr	5
112	raw, weighed with stones	0	90	0	(0.67)	0.05	0.04	0.6	0.1	(0.05)	0	(3)	(0.19)	Tr	5
113	stewed with sugar	0	85	0	(0.63)	0.03	0.03	0.4	0.1	(0.04)	0	Tr	(0.14)	Tr	3
114	stewed with sugar, weighed with stones	0	82	0	(0.60)	0.03	0.03	0.4	0.1	(0.04)	0	Tr	(0.13)	Tr	3
115	stewed without sugar	0	85	0	(0.63)	0.03	0.03	0.4	0.1	(0.04)	0	Tr	(0.18)	Tr	3
116	stewed without sugar, weighed with stones	0	81	0	(0.60)	0.03	0.03	0.4	0.1	(0.03)	0	Tr	(0.17)	Tr	3
117	**Grenadillas**	0	10	0	N	Tr	0.10	1.9	0.5	N	0	N	N	N	20
118	**Guava**, raw	0	435[a]	0	N	0.04	0.04	1.0	0.1	0.14	0	N	0.15	N	230[b]
119	raw, weighed with skin and pips	0	390	0	N	0.04	0.04	0.9	0.1	0.13	0	N	0.13	N	210
120	canned in syrup	0	(145)	0	N	(0.02)	(0.02)	(0.6)	0.1	(0.09)	0	N	(0.09)	N	180
121	**Jambu fruit**	0	N	0	N	0.01	0.02	0.4	N	N	0	N	N	N	12
122	**Jujube**	0	24	0	N	0.03	0.03	0.8	0.2	0.08	0	N	N	N	57
123	**Kiwi fruit**	0	37	0	N	0.01	0.03	0.3	0.3	0.15	0	N	N	N	59
124	weighed with skin	0	32	0	N	0.01	0.03	0.3	0.3	0.13	0	N	N	N	51

a Peel included on analysis

b Levels ranged from 9 to 410mg vitamin C per 100g

Fruit *continued*

Composition of food per 100g

No. 14-	Food	Description and main data sources	Edible proportion	Water g	Total nitrogen g	Protein g	Fat g	Carbo-hydrate g	Energy value kcal	kJ
125	**Kumquats**, *raw*	Analysis and literature sources; whole fruit	1.00	83.4	0.14	0.9	0.5	9.3	43	183
126	canned in syrup	Ref. 11	1.00	61.7	0.06	0.4	0.5	35.4	138	577
127	**Lemon peel**	Ref. 3	1.00	81.6	0.24	1.5	0.3	N	N	N
128	**Lemons**, *whole, without pips*	Analysis and literature sources; includes peel but no pips	0.99	86.3	0.16	1.0	0.3	3.2	19	79
129	*peeled*	Literature sources; flesh only	1.00	89.1	0.13	0.8	0.3	2.2	14	60
130	*peeled, raw, weighed with peel and pips*	Calculated from 129	0.64	57.0	0.08	0.5	0.2	1.4	9	38
131	**Limes**, *peeled*	Literature sources; flesh only	1.00	89.7	0.11	0.7	0.3	0.8	9	36
132	*peeled, weighed with peel and pips*	Calculated from 131	0.74	66.4	0.08	0.5	0.2	0.6	6	25
133	**Loganberries**, *raw*	Whole fruit	1.00	85.0	0.17	1.1	Tr	3.4	17	73
134	*stewed with sugar*	Calculated from 700g fruit, 210g water, 84g sugar	1.00	78.9	0.13	0.8	Tr	12.5	50	214
135	*stewed without sugar*	Calculated from 700g fruit, 210g water	1.00	87.2	0.15	0.9	Tr	2.9	14	62
136	*canned in juice*	Fruit and syrup	1.00	66.3	0.10	0.6	Tr	26.2	101	429
137	**Longans**, *dried*	Literature sources; flesh only	1.00	17.6	0.78	4.9	0.4	72.0	286	1197
138	*canned in syrup, drained*	Ref. 11; fruit only	0.45	81.4	0.06	0.4	0.3	17.1	67	276
139	*whole contents*	Ref. 11; fruit and syrup	1.00	82.5	0.05	0.3	0.2	16.5	63	264
140	**Loquats**, *raw*	Literature sources; flesh only	0.63	88.5	0.11	0.7	0.2	6.3	28	115
141	*canned in syrup*	Ref. 11	1.00	76.7	0.05	0.3	0.1	22.2	84	351

Fruit *continued*

Carbohydrate fractions, g per 100g

No. 14-	Food	Starch	Total sugars	Individual sugars					Dietary fibre		Fibre fractions			
				Gluc	Fruct	Sucr	Malt	Lact	Southgate method	Englyst method	Cellulose	Non-cellulosic polysaccharide Soluble	Non-cellulosic polysaccharide Insoluble	Lignin
125	**Kumquats**, *raw*	0	9.3	2.6	3.1	3.6	0	0	N	3.8	0.8	2.4	0.5	N
126	*canned in syrup*	0	35.4	N	N	N	0	0	N	1.7	0.4	1.1	0.2	N
127	**Lemon peel**	0	N	N	N	N	0	0	N	N	N	N	N	N
128	**Lemons**, *whole, without pips*	0	3.2	1.4	1.4	0.4	0	0	N	N	N	N	N	N
129	*peeled*	0	2.2	0.9	0.7	0.5	0	0	4.7	N	N	N	N	Tr
130	*peeled, raw, weighed with peel and pips*	0	1.4	0.6	0.4	0.3	0	0	N	N	N	N	N	Tr
131	**Limes**, *peeled*	0	0.8	0.3	0.3	0.1	0	0	N	N	N	N	N	Tr
132	*peeled, weighed with peel and pips*	0	0.6	0.2	0.2	0.1	0	0	N	N	N	N	N	Tr
133	**Loganberries**, *raw*	0	3.4	1.9	1.3	0.2	0	0	N	(2.5)	(1.2)	(0.7)	(0.6)	Tr
134	*stewed with sugar*	0	12.5	2.0	1.5	9.0	0	0	5.6	2.0	0.9	0.5	0.5	N
135	*stewed without sugar*	0	2.9	1.6	1.1	0.2	0	0	4.4	2.1	1.0	0.6	0.5	N
136	*canned in juice*	0	26.2	N	N	N	0	0	4.8	1.6	0.8	0.3	0.5	N
137	**Longans**, *dried*	0.5	71.5	N	N	N	0	0	3.0	N	N	N	N	0.1
138	*canned in syrup, drained*	0.1	17.0	N	N	N	0	0	4.4	N	N	N	N	Tr
139	*whole contents*	0.1	16.4	N	N	N	0	0	1.0	N	N	N	N	Tr
140	**Loquats**, *raw*	0	6.3	2.7	3.0	0.6	0	0	0.5	N	N	N	N	N
141	*canned in syrup*	0	22.2	N	N	N	0	0	N	N	N	N	N	N

Fruit continued

Inorganic constituents per 100g

No. 14-	Food	Na	K	Ca	Mg	P	Fe	Cu	Zn	S	Cl	Mn	Se	I
						mg							µg	
125	**Kumquats**, *raw*	6	180	25	13	49	0.6	0.11	0.1	N	N	0.1	N	N
126	canned in syrup	(5)	160	16	N	65	0.8	N	N	N	N	N	N	N
127	**Lemon peel**	6	160	130	15	12	0.8	N	N	N	N	N	N	N
128	**Lemons**, *whole, without pips*	5	150	85	12	18	0.5	0.26	0.1	12	5	N	(1)	N
129	*peeled*	3	140	27	9	16	0.4	Tr	0.1	N	N	N	(1)	N
130	*peeled, raw, weighed with peel and pips*	2	90	17	6	10	0.3	Tr	0.1	N	N	N	(1)	N
131	**Limes**, *peeled*	2	130	23	11	18	0.4	0.05	0.1	N	N	Tr	N	N
132	*peeled, weighed with peel and pips*	1	96	17	8	13	0.3	0.04	0.1	N	N	Tr	N	N
133	**Loganberries**, *raw*	3	260	35	25	24	1.4	0.14	0.3	18	16	0.9	Tr	N
134	*stewed with sugar*	2	200	27	19	18	1.1	0.11	0.3	14	12	0.7	Tr	N
135	*stewed without sugar*	2	220	29	21	20	1.2	0.12	0.3	15	13	0.8	Tr	N
136	*canned in juice*	1	97	18	11	23	1.4	0.04	N	3	5	N	Tr	N
137	**Longans**, *dried*	48	660	45	46	200	5.4	0.81	0.2	N	N	0.3	N	N
138	*canned in syrup, drained*	53	40	7	N	8	0.7	N	N	N	N	N	N	N
139	*whole contents*	41	110	5	N	8	0.3	N	N	N	N	N	N	N
140	**Loquats**, *raw*	1	220	20	10	22	0.4	0.04	0.2	N	N	0.1	1	N
141	*canned in syrup*	N	N	22	N	3	0.1	N	N	N	N	N	Tr	N

No. 14-	Food	Retinol µg	Carotene µg	Vitamin D µg	Vitamin E mg	Thiamin mg	Ribo-flavin mg	Niacin mg	Trypt 60 mg	Vitamin B6 mg	Vitamin B12 µg	Folate µg	Panto-thenate mg	Biotin µg	Vitamin C mg
125	**Kumquats**, raw	0	175	0	N	0.07	0.07	0.5	0.1	N	0	N	N	N	39
126	canned in syrup	0	(25)	0	N	(0.02)	(0.02)	(0.2)	Tr	N	0	N	N	N	(10)
127	**Lemon peel**	0	30	0	N	0.06	0.08	0.4	0.2	0.17	0	N	0.32	N	130
128	**Lemons**, whole, without pips	0	18	0	N	0.05	0.04	0.2	0.1	0.11	0	N	0.23	0.5	58
129	peeled	0	7	0	N	0.05	0.02	0.2	0.1	0.08	0	11	0.19	N	53
130	peeled, raw, weighed with peel and pips	0	4	0	N	0.03	0.01	0.1	0.1	0.05	0	7	0.12	N	34
131	**Limes**, peeled	0	12	0	N	0.03	0.02	0.2	0.1	(0.08)	0	8	0.22	N	46
132	peeled, weighed with peel and pips	0	9	0	N	0.02	0.01	0.1	0.1	(0.06)	0	6	0.16	N	34
133	**Loganberries**, raw	0	(6)	0	(0.48)	(0.03)	(0.05)	(0.5)	0.2	(0.06)	0	(33)	(0.24)	(1.9)	35
134	stewed with sugar	0	(4)	0	(0.38)	(0.02)	(0.03)	(0.3)	0.2	(0.04)	0	(5)	(0.14)	(1.1)	21
135	stewed without sugar	0	(5)	0	(0.41)	(0.02)	(0.03)	(0.3)	0.2	(0.04)	0	(5)	(0.15)	(1.2)	22
136	canned in juice	0	(3)	0	(0.26)	(0.01)	0.02	(0.3)	0.1	(0.04)	0	(11)	(0.17)	(0.7)	25
137	**Longans**, dried	0	0	0	N	0.04	(0.50)	(1.0)	N	N	0	N	N	N	28
138	canned in syrup, drained	0	0	0	N	Tr	0.08	0.1	N	N	0	N	N	N	68
139	whole contents	0	0	0	N	Tr	0.04	0.1	N	N	0	N	N	N	63
140	**Loquats**, raw	0	515	0	N	0.02	0.03	0.2	N	N	0	N	N	N	3
141	canned in syrup	0	95	0	N	0.01	Tr	0.5	N	N	0	N	N	N	0

45

Fruit continued

No. 14-	Food	Description and main data sources	Edible proportion	Water g	Total nitrogen g	Protein g	Fat g	Carbo-hydrate g	Energy value kcal	kJ
142	**Lychees**, *raw*	Analysis and literature sources; flesh only	1.00	81.1	0.14	0.9	0.1	14.3	58	248
143	*raw, weighed with skin and stone*	Calculated from *142*	0.62	50.3	0.09	0.5	0.1	8.9	36	155
144	*canned in syrup*	Analysis and literature sources	1.00	79.3	0.06	0.4	Tr	17.7	68	290
145	**Mammie apple**	Literature sources; flesh only	0.54	86.4	0.08	0.5	0.4	14.4	60	254
146	**Mandarin oranges**, canned in juice	10 samples, 4 brands	1.00	89.6	0.11	0.7	Tr	7.7	32	135
147	*canned in syrup*	10 samples, 10 brands	1.00	84.8	0.08	0.5	Tr	13.4	52	223
148	**Mangoes**, *ripe, raw*	Literature sources; flesh only	1.00	82.4	0.11	0.7	0.2	14.1	57	245
149	*raw, weighed with skin and stone*	Calculated from *148*	0.68	56.0	0.07	0.5	0.1	9.6	39	166
150	*canned in syrup*	10 samples	1.00	74.8	0.05	0.3	Tr	20.3	77	330
151	**Mangosteen**	Literature sources; flesh only	0.24	81.0	0.10	0.6	0.5	16.4	73	307
152	**Medlars**	Flesh only	0.81	74.5	0.08	0.5	Tr	10.6	42	178

Fruit *continued*

Carbohydrate fractions, g per 100g

No. 14-	Food	Starch	Total sugars	Gluc	Fruct	Sucr	Malt	Lact	Dietary fibre Southgate method	Dietary fibre Englyst method	Cellulose	Non-cellulosic polysaccharide Soluble	Non-cellulosic polysaccharide Insoluble	Lignin
142	**Lychees**, *raw*	0	14.3	7.0	7.3	Tr	0	0	1.5	0.7	0.1	0.5	0.1	Tr
143	*raw, weighed with skin and stone*	0	8.9	4.3	4.5	Tr	0	0	0.9	0.4	0.1	0.3	0.1	Tr
144	canned in syrup	0	17.7	8.5	8.6	0.6	0	0	0.7	0.5	0.1	0.3	0.1	Tr
145	**Mammie apple**	Tr	14.4	N	N	N	0	0	N	N	N	N	N	N
146	**Mandarin oranges**, canned in juice	0	7.7	2.8	3.1	1.8	0	0	(0.3)	0.3	Tr	0.2	0.1	Tr
147	canned in syrup	0	13.4	4.1	4.2	5.1	0	0	0.3	0.2	Tr	0.1	0.1	Tr
148	**Mangoes**, *ripe, raw*	0.3	13.8	0.7	3.0	10.1	0	0	(2.9)	2.6	0.5	1.6	0.5	0.3
149	*raw, weighed with skin and stone*	0.2	9.4	0.5	2.0	6.9	0	0	(2.0)	1.8	0.3	1.1	0.3	0.2
150	canned in syrup	0.1	20.2	7.0	8.8	4.4	0	0	0.9	0.7	0.3	0.4	Tr	Tr
151	**Mangosteen**	0.3	16.1	N	N	N	0	0	1.3	N	N	N	N	Tr
152	**Medlars**	0	10.6	N	N	N	0	0	9.2	N	N	N	N	N

Fruit *continued*

Inorganic constituents per 100g

No. 14-	Food	Na	K	Ca	Mg	P	Fe	Cu	Zn	S	Cl	Mn	Se	I
		mg											µg	
142	**Lychees**, *raw*	1	160	6	9	30	0.5	0.15	0.3	19	3	0.1	N	N
143	*raw, weighed with skin and stone*	1	99	4	6	19	0.3	0.09	0.2	12	2	Tr	N	N
144	canned in syrup	2	75	4	6	12	0.7	0.11	0.2	N	(5)	N	N	N
145	**Mammie apple**	15	47	12	N	11	0.6	N	N	N	N	N	N	N
146	**Mandarin oranges**, canned in juice	6	85	17	9	13	0.5	Tr	0.1	N	2	Tr	Tr	Tr
147	canned in syrup	6	49	17	7	8	0.2	Tr	Tr	N	2	Tr	Tr	Tr
148	**Mangoes**, *ripe, raw*	2	180	12	13	16	0.7	0.12	0.1	N	N	0.3	N	N
149	*raw, weighed with skin and stone*	1	120	8	9	11	0.5	0.08	0.1	N	N	0.2	N	N
150	canned in syrup	3	100	10	7	10	0.4	0.09	0.3	N	(5)	N	N	N
151	**Mangosteen**	1	130	14	N	14	0.3	N	N	N	N	N	N	N
152	**Medlars**	6	250	30	11	28	0.5	0.17	N	17	3	N	N	N

Fruit *continued*

No. 14-	Food	Retinol µg	Carotene µg	Vitamin D µg	Vitamin E mg	Thiamin mg	Ribo-flavin mg	Niacin mg	Trypt 60 mg	Vitamin B6 mg	Vitamin B12 µg	Folate µg	Panto-thenate mg	Biotin µg	Vitamin C mg
142	**Lychees**, *raw*	0	0	0	N	0.04	0.06	0.5	0.1	N	0	N	N	N	45
143	*raw, weighed with skin and stone*	0	0	0	N	0.02	0.04	0.3	Tr	N	0	N	N	N	28
144	*canned in syrup*	0	0	0	N	Tr	0.04	Tr	0.1	N	0	N	N	N	8
145	**Mammie apple**	0	(130)	0	N	0.02	0.04	0.4	0.1	N	0	N	0.10	N	15
146	**Mandarin oranges**, canned in														
	juice	0	95	0	Tr	0.08	0.01	0.2	0.1	(0.03)	0	12	(0.15)	(0.8)	20
147	*canned in syrup*	0	105	0	Tr	0.06	0.01	0.2	Tr	0.03	0	12	(0.15)	(0.8)	15
148	**Mangoes**, ripe, *raw*	0	1800[a]	0	1.05	0.04	0.05	0.5	1.3	0.13	0	N	0.16	N	37
149	*raw, weighed with skin and stone*	0	1225	0	0.71	0.03	0.03	0.3	0.1	0.09	0	N	0.11	N	25
150	*canned in syrup*	0	1470	0	0.64	(0.02)	(0.03)	(0.2)	0.1	(0.07)	0	N	(0.04)	N	10
151	**Mangosteen**	0	0	0	N	0.05	0.01	0.3	0.1	N	0	N	N	N	3
152	**Medlars**	0	N	0	N	N	N	N	0.1	N	0	N	N	N	2

[a] Levels ranged from 300 to 3000µg carotene per 100g

Composition of food per 100g

No. 14-	Food	Description and main data sources	Edible proportion	Water g	Total nitrogen g	Protein g	Fat g	Carbohydrate g	Energy value kcal	kJ
153	**Melon,** *average*	Average of Canteloupe-type, Galia and Honeydew varieties; flesh only	1.00	92.0	0.09	0.6	0.1	5.5	24	102
154	*average, weighed whole*	Calculated from 153	0.62	57.0	0.06	0.4	0.1	3.4	15	65
155	*-, weighed with skin*	Calculated from 153; no pips	0.66	61.4	0.06	0.4	0.1	3.6	16	68
156	Canteloupe-type	10 samples, Canteloupe, Charantais and Rock; flesh only	1.00	92.1	0.10	0.6	0.1	4.2	19	81
157	*weighed whole*	Calculated from 156	0.59	54.3	0.06	0.4	0.1	2.5	12	51
158	*weighed with skin*	Calculated from 156; no pips	0.66	60.8	0.07	0.4	0.1	2.8	13	55
159	Galia	11 samples; flesh only	1.00	91.7	0.08	0.5	0.1	5.6	24	102
160	*weighed whole*	Calculated from 159	0.64	58.7	0.05	0.3	0.1	3.6	16	66
161	*weighed with skin*	Calculated from 159; no pips	0.68	63.4	0.05	0.3	0.1	3.8	16	70
162	Honeydew	10 samples; flesh only	1.00	92.2	0.10	0.6	0.1	6.6	28	119
163	*weighed whole*	Calculated from 162	0.63	58.1	0.06	0.4	0.1	4.2	18	78
164	*weighed with skin*	Calculated from 162; no pips	0.65	59.9	0.07	0.4	0.1	4.3	19	77
165	**watermelon**	Literature sources; flesh only	1.00	92.3	0.07	0.5	0.3	7.1	31	133
166	*weighed whole*	Calculated from 165	0.57	52.6	0.04	0.3	0.2	4.0	18	77
167	**Mixed peel**	10 samples, 9 brands	1.00	20.9	0.05	0.3	0.9	59.1	231	984
168	**Mulberries,** *raw*	Whole fruit	1.00	85.0	0.21	1.3	Tr	8.1	36	152
169	*stewed with sugar*	Calculated from 700g fruit, 210g water, 84g sugar	1.00	78.9	0.16	1.0	Tr	16.2	65	276
170	*stewed without sugar*	Calculated from 700g fruit, 210g water	1.00	87.2	0.18	1.1	Tr	6.9	30	129

No. 14-	Food	Starch	Total sugars	Individual sugars					Dietary fibre		Fibre fractions			
				Gluc	Fruct	Sucr	Malt	Lact	Southgate method	Englyst method	Cellulose	Non-cellulosic polysaccharide Soluble	Insoluble	Lignin
153	**Melon**, average	0	5.5	2.1	2.5	0.9	0	0	0.9	0.7	0.3	0.2	0.1	Tr
154	average, weighed whole	0	3.4	1.3	1.5	0.6	0	0	0.6	0.4	0.2	0.1	0.1	Tr
155	-, weighed with skin	0	3.6	1.4	1.7	0.6	0	0	0.6	0.5	0.2	0.1	0.1	Tr
156	Canteloupe-type	0	4.2	1.8	2.2	0.1	0	0	0.9	1.0	0.5	0.3	0.2	Tr
157	weighed whole	0	2.5	1.1	1.3	0.1	0	0	0.5	0.6	0.3	0.2	0.1	Tr
158	weighed with skin	0	2.8	1.2	1.5	0.1	0	0	0.6	0.7	0.3	0.2	0.1	Tr
159	Galia	0	5.6	1.6	2.0	2.0	0	0	(0.9)	0.4	0.2	0.1	0.1	Tr
160	weighed whole	0	3.6	1.0	1.3	1.3	0	0	(0.6)	0.3	0.1	0.1	0.1	Tr
161	weighed with skin	0	3.8	1.1	1.4	1.4	0	0	(0.6)	0.3	0.1	0.1	0.1	Tr
162	Honeydew	0	6.6	2.8	3.2	0.6	0	0	0.8	0.6	0.3	0.2	0.1	Tr
163	weighed whole	0	4.2	1.8	2.0	0.4	0	0	0.5	0.4	0.2	0.1	0.1	Tr
164	weighed with skin	0	4.3	1.8	2.1	0.4	0	0	0.5	0.4	0.2	0.1	0.1	Tr
165	**watermelon**	0	7.1	1.3	2.3	3.4	0	0	0.3	0.1	Tr	0.1	Tr	Tr
166	weighed whole	0	4.0	0.7	1.3	1.9	0	0	0.2	0.1	Tr	0.1	Tr	Tr
167	**Mixed peel**	0	59.1	19.9	4.5	9.1	25.6	0	N	4.8	1.1	2.9	0.8	N
168	**Mulberries**, raw	0	8.1	3.8	4.3	Tr	0	0	1.5	N	N	N	N	N
169	stewed with sugar	0	16.2	3.5	3.9	8.9	0	0	1.2	N	N	N	N	N
170	stewed without sugar	0	6.9	3.3	3.7	Tr	0	0	1.3	N	N	N	N	N

Fruit continued

No. 14-	Food	Na	K	Ca	Mg	P	Fe	Cu	Zn	S	Cl	Mn	Se	I
							mg						μg	
153	**Melon, average**	24	190	14	11	13	0.2	Tr	Tr	9	55	Tr	Tr	N
154	average, weighed whole	15	120	9	7	8	0.1	Tr	Tr	6	34	Tr	Tr	N
155	-, weighed with skin	16	130	7	7	9	0.1	Tr	Tr	6	36	Tr	Tr	N
156	Canteloupe-type	8	210	20	11	13	0.3	Tr	Tr	12	44	Tr	Tr	(4)
157	weighed whole	5	120	12	6	8	0.2	Tr	Tr	7	26	Tr	Tr	(2)
158	weighed with skin	5	140	13	7	9	0.2	Tr	Tr	8	29	Tr	Tr	(3)
159	Galia	31	150	13	12	10	0.2	Tr	0.1	9	75	Tr	Tr	N
160	weighed whole	20	96	8	8	6	0.1	Tr	0.1	6	48	Tr	Tr	N
161	weighed with skin	21	100	9	8	7	0.1	Tr	0.1	6	51	Tr	Tr	N
162	Honeydew	32	210	9	10	16	0.1	Tr	Tr	6	45	Tr	Tr	N
163	weighed whole	20	130	6	6	10	0.1	Tr	Tr	4	28	Tr	Tr	N
164	weighed with skin	21	140	6	7	10	0.1	Tr	Tr	4	29	Tr	Tr	N
165	**watermelon**	2	100	7	8	9	0.3	0.03	0.2	N	N	Tr	Tr	Tr
166	weighed whole	1	57	4	5	5	0.2	0.02	0.1	N	N	Tr	Tr	Tr
167	**Mixed peel**	280	21	130	12	6	1.3	0.15	0.2	N	N	0.1	N	N
168	**Mulberries,** raw	2	260	36	15	48	1.6	0.06	0.2	9	4	(0.9)	Tr	N
169	stewed with sugar	1	200	28	11	37	1.3	0.05	0.2	7	3	(0.7)	Tr	N
170	stewed without sugar	1	220	30	12	41	1.4	0.05	0.2	7	3	(0.8)	Tr	N

Fruit continued

No. 14-	Food	Retinol µg	Carotene µg	Vitamin D µg	Vitamin E mg	Thiamin mg	Ribo-flavin mg	Niacin mg	Trypt 60 mg	Vitamin B6 mg	Vitamin B12 µg	Folate µg	Panto-thenate mg	Biotin µg	Vitamin C mg
153	**Melon**, average	0	N	0	0.10	0.03	0.01	0.4	Tr	0.09	0	3	0.17	N	17
154	average, weighed whole	0	N	0	0.06	0.02	0.01	0.2	Tr	0.06	0	2	0.11	N	11
155	-, weighed with skin	0	N	0	0.08	0.02	0.01	0.3	Tr	0.06	0	2	0.12	N	11
156	Cantaloupe-type	0	1000[a]	0	0.10	0.04	0.02	0.6	Tr	0.11	0	5	0.13	N	26
157	weighed whole	0	590	0	0.06	0.02	0.01	0.3	Tr	0.06	0	3	0.08	N	15
158	weighed with skin	0	660	0	0.10	0.03	0.01	0.4	Tr	0.07	0	3	0.09	N	17
159	Galia	0	N	0	(0.10)	(0.03)	(0.01)	(0.4)	Tr	(0.09)	0	(3)	(0.17)	N	15
160	weighed whole	0	N	0	(0.06)	(0.02)	(0.01)	(0.3)	Tr	(0.06)	0	(2)	(0.11)	N	10
161	weighed with skin	0	N	0	(0.07)	(0.02)	(0.01)	(0.3)	Tr	(0.06)	0	(2)	(0.12)	N	10
162	Honeydew	0	48	0	0.10	0.03	0.01	0.3	Tr	0.06	0	2	0.21	N	9
163	weighed whole	0	30	0	0.06	0.02	0.01	0.2	Tr	0.04	0	1	0.13	N	6
164	weighed with skin	0	31	0	0.07	0.02	0.01	0.2	Tr	0.04	0	1	0.14	N	6
165	**watermelon**	0	230	0	(0.10)	0.05	0.01	0.1	Tr	0.14	0	2	0.21	1.0	8
166	weighed whole	0	130	0	(0.06)	0.03	0.01	0.1	Tr	0.08	0	1	0.12	0.6	5
167	**Mixed peel**	0	Tr	0	N	N	N	N	0.1	N	0	N	N	N	Tr
168	**Mulberries**, raw	0	14	0	N	0.03	0.05	0.7	0.2	(0.06)	0	(33)	(0.24)	(1.9)	19
169	stewed with sugar	0	10	0	N	0.02	0.03	0.4	0.2	(0.04)	0	(5)	(0.14)	(1.1)	11
170	stewed without sugar	0	11	0	N	0.02	0.03	0.5	0.2	(0.04)	0	(5)	(0.15)	(1.2)	12

[a] This is an average value. Carotene levels have been reported for Rock melons at 835µg and for Canteloupe melons at 1510 to 1930µg per 100g

Composition of food per 100g

No. 14-	Food	Description and main data sources	Edible proportion	Water g	Total nitrogen g	Protein g	Fat g	Carbo-hydrate g	Energy value kcal	kJ
171	**Nectarines**	10 samples; flesh and skin	1.00	88.9	0.22	1.4	0.1	9.0	40	171
172	*weighed with stones*	Calculated from *171*	0.89	79.1	0.20	1.2	0.1	8.0	36	152
173	**Olives,** *in brine*	Bottled, drained; flesh and skin, green	1.00	76.5	0.14	0.9	11.0	Tr	103	422
174	*in brine, weighed with stones*	Calculated from *173*	0.80	61.2	0.11	0.7	8.8	Tr	82	337
175	**Oranges**	Assorted varieties; flesh only	1.00	86.1	0.18	1.1	0.1	8.5	37	158
176	*weighed with peel and pips*	Calculated from *175*	0.70[a]	60.3	0.13	0.8	0.1	5.9	26	112
177	**Ortaniques**	Ref. 1; flesh only	0.73	86.0	0.16	1.0	0.2	11.7	49	205
178	**Passion fruit**	Analysis and literature sources; flesh and pips	1.00	74.9	0.45	2.6	0.4	5.8	36	152
179	*weighed with skin*	Calculated from *178*	0.61	45.7	0.27	1.7	0.2	3.5	22	92
180	**Paw-paw,** *raw*	Literature sources; flesh only	1.00	88.5	0.08	0.5	0.1	8.8	36	153
181	*raw, weighed with skin and pips*	Calculated from *180*	0.75	66.4	0.06	0.4	0.1	6.6	27	116
182	*canned in juice*	10 samples	1.00	80.4	0.03	0.2	Tr	17.0	65	275
183	**Peaches,** *raw*	10 samples; flesh and skin	1.00	88.9	0.16	1.0	0.1	7.6	33	142
184	*raw, weighed with stone*	Calculated from *183*	0.90	80.0	0.14	0.9	0.1	6.8	30	128
185	*dried*	No stones	1.00	15.5	0.55	3.4	0.8	53.0	219	936
186	*dried, stewed with sugar*	Calculated from 450g fruit, 770g water, 54g sugar	1.00	62.1	0.22	1.3	0.3	25.7	104	446
187	*-, stewed without sugar*	Calculated from 450g fruit, 770g water	1.00	65.4	0.23	1.4	0.3	21.7	89	383
188	*canned in juice*	10 samples, 7 brands; halves and slices	1.00	86.7	0.09	0.6	Tr	9.7	39	165
189	*canned in syrup*	10 samples, 9 brands; halves and slices	1.00	81.1	0.08	0.5	Tr	14.0	55	233

[a] Levels ranged from 0.60 to 0.74

Carbohydrate fractions, g per 100g

No. Food 14-	Starch	Total sugars	Individual sugars Gluc	Fruct	Sucr	Malt	Lact	Dietary fibre Southgate method	Englyst method	Fibre fractions Cellulose	Non-cellulosic polysaccharide Soluble	Insoluble	Lignin
171 **Nectarines**	0	9.0	1.3	1.3	6.3	0	0	2.2	1.2	0.4	0.6	0.2	0.2
172 *weighed with stones*	0	8.0	1.2	1.2	5.6	0	0	2.0	1.1	0.4	0.5	0.2	0.2
173 **Olives,** *in brine*	0	Tr	Tr	Tr	Tr	0	0	4.0	2.9	0.9	0.3	1.7	Tr
174 *in brine, weighed with stones*	0	Tr	Tr	Tr	Tr	0	0	3.2	2.3	0.7	0.2	1.4	Tr
175 **Oranges**	0	8.5	2.2	2.4	3.9	0	0	1.8	1.7	0.4	1.1	0.2	0.1
176 *weighed with peel and pips*	0	5.9	1.5	1.7	2.7	0	0	1.3	1.2	0.3	0.8	0.1	0.1
177 **Ortaniques**	0	11.7	N	N	N	0	0	(1.8)	(1.7)	(0.4)	(1.1)	(0.2)	(0.1)
178 **Passion fruit**	0	5.8	2.2	1.9	1.7	0	0	N	3.3	1.4	0.5	1.4	N
179 *weighed with skin*	0	3.5	1.3	1.1	1.0	0	0	N	2.0	0.9	0.3	0.9	N
180 **Paw-paw,** *raw*	0	8.8	2.8	2.8	3.1	0	0	2.3	2.2	0.7	1.3	0.2	0.1
181 *raw, weighed with skin and pips*	0	6.6	2.1	2.1	2.3	0	0	1.7	1.7	0.5	1.0	0.1	0.1
182 *canned in juice*	0	17.0	7.8	7.0	2.2	0	0	(0.7)	0.7	0.4	0.3	0.1	Tr
183 **Peaches,** *raw*	0	7.6	1.1	1.1	5.2	0	0	2.3	1.5	0.5	0.8	0.2	0.1
184 *raw, weighed with stone*	0	6.8	1.0	1.0	4.7	0	0	2.1	1.3	0.5	0.7	0.2	0.1
185 *dried*	0	53.0	(7.9)	(7.9)	(37.1)	0	0	12.9	7.3	2.1	3.8	1.4	N
186 *dried, stewed with sugar*	0	25.7	(4.0)	(4.0)	(17.6)	0	0	5.1	2.9	0.8	1.5	0.5	N
187 *-, stewed without sugar*	0	21.7	(3.9)	(3.9)	(13.7)	0	0	5.3	3.0	0.9	1.6	0.6	N
188 *canned in juice*	0	9.7	2.4	3.7	3.6	0	0	(0.9)	0.8	0.3	0.4	0.1	Tr
189 *canned in syrup*	0	14.0	3.7	3.6	6.7	0	0	0.9	0.9	0.3	0.5	0.1	Tr

Fruit continued

Inorganic constituents per 100g

No. 14-	Food	Na	K	Ca	Mg	P	Fe (mg)	Cu	Zn	S	Cl	Mn	Se (µg)	I
171	**Nectarines**	1	170	7	10	22	0.4	0.06	0.1	10	5	0.1	(1)	3
172	weighed with stones	1	150	6	9	20	0.4	0.05	0.1	9	4	0.1	(1)	3
173	**Olives**, in brine	2250	91	61	22	17	1.0	0.23	N	36	3750	N	N	N
174	in brine, weighed with stones	1800	73	49	18	14	0.8	0.18	N	29	3000	N	N	N
175	**Oranges**	5	150	47	10	21	0.1	0.05	0.1	10	3	Tr	(1)	2
176	weighed with peel and pips	3	110	33	7	15	0.1	0.03	0.1	7	2	Tr	(1)	1
177	**Ortaniques**	(5)	(150)	41	(10)	(21)	0.4	(0.05)	(0.1)	(10)	(3)	Tr	N	(2)
178	**Passion fruit**	19	200	11	29	64	1.3	N	0.8	N	N	N	N	N
179	weighed with skin	12	120	7	18	39	0.8	N	0.5	N	N	N	N	N
180	**Paw-paw**, raw	5	200	23	11	13	0.5	0.08	0.2	13	11	0.1	N	3
181	raw, weighed with skin and pips	4	170	17	8	10	0.4	0.06	0.1	10	8	0.1	N	N
182	canned in juice	8	110	23	8	6	0.4	0.10	0.3	N	40	N	N	N
183	**Peaches**, raw	1	160	7	9	22	0.4	0.06	0.1	6	Tr	0.1	(1)	3
184	raw, weighed with stone	1	140	6	8	20	0.4	0.05	0.1	5	Tr	0.1	(1)	3
185	dried	6	1100	36	54	120	6.8	0.63	0.8	240	11	0.8	(8)	23
186	dried, stewed with sugar	2	430	14	21	47	2.7	0.25	0.3	94	4	0.3	(3)	9
187	-, stewed without sugar	2	450	14	22	49	2.8	0.26	0.3	98	4	0.3	(3)	9
188	canned in juice	12	170	4	7	19	0.4	0.04	0.1	(1)	(4)	0.1	Tr	N
189	canned in syrup	4	110	3	5	11	0.2	Tr	Tr	1	4	Tr	Tr	N

No. 14-	Food	Retinol μg	Carotene μg	Vitamin D μg	Vitamin E mg	Thiamin mg	Riboflavin mg	Niacin mg	Trypt 60 mg	Vitamin B6 mg	Vitamin B12 μg	Folate μg	Pantothenate mg	Biotin μg	Vitamin C mg
171	**Nectarines**	0	58	0	N	0.02	0.04	0.6	0.3	0.03	0	Tr	0.16	(0.2)	37
172	weighed with stones	0	52	0	N	0.02	0.04	0.5	0.3	0.03	0	Tr	0.14	(0.2)	33
173	**Olives**, in brine	0	180[a]	0	1.99	Tr	Tr	Tr	0.1	0.02	0	Tr	0.02	Tr	0
174	in brine, weighed with stones	0	145	0	1.59	Tr	Tr	Tr	0.1	0.02	0	Tr	0.02	Tr	0
175	**Oranges**	0	28[b]	0	0.24	0.11	0.04	0.4	0.1	0.10	0	31	0.37	1.0	54[c]
176	weighed with peel and pips	0	20	0	0.17	0.08	0.03	0.3	0.1	0.07	0	22	0.26	0.7	38
177	**Ortaniques**	0	(120)	0	(0.24)	0.10	0.04	0.4	0.1	(0.10)	0	(31)	(0.37)	(1.0)	50
178	**Passion fruit**	0	750	0	N	0.03	0.12	1.5	0.4	N	0	N	N	N	23
179	weighed with skin	0	460	0	N	0.02	0.07	0.9	0.2	N	0	N	N	N	14
180	**Paw-paw**, raw	0	810	0	N	0.03	0.04	0.3	0.1	0.03	0	1	0.22	N	60
181	raw, weighed with skin and pips	0	585	0	N	0.02	0.03	0.2	0.1	0.02	0	1	0.17	N	45
182	canned in juice	0	(255)	0	N	0.02	0.02	0.2	Tr	(0.01)	0	Tr	(0.20)	N	15
183	**Peaches**, raw	0	58	0	N	0.02	0.04	0.6	0.2	0.02	0	3	0.17	(0.2)	31
184	raw, weighed with stone	0	53	0	N	0.02	0.04	0.5	0.2	0.02	0	3	0.15	(0.2)	28
185	dried	0	445	0	N	Tr	0.19	5.3	0.7	0.10	0	(14)	(0.30)	N	Tr
186	dried, stewed with sugar	0	175	0	N	Tr	0.06	1.6	0.3	0.03	0	(1)	(0.09)	N	Tr
187	-, stewed without sugar	0	180	0	N	Tr	0.06	1.6	0.3	0.03	0	(1)	(0.09)	N	Tr
188	canned in juice	0	67	0	N	0.01	0.01	0.6	0.1	0.02	0	2	0.06	0.2	6
189	canned in syrup	0	75	0	N	0.01	0.01	0.6	0.1	0.02	0	7	0.05	0.1	5

[a] Values for green olives. Ripe black olives contain 40μg carotene per 100g
[b] Blood oranges have been found to contain 155μg carotene per 100g
[c] Levels ranged from 44 to 79mg vitamin C per 100g

Fruit *continued*

Composition of food per 100g

No. 14-	Food	Description and main data sources	Edible proportion	Water g	Total nitrogen g	Protein g	Fat g	Carbo-hydrate g	Energy value kcal	kJ
190	**Pears,** *average, raw*	Average of Comice, Conference and Williams varieties; flesh and skin	1.00	83.8	0.05	0.3	0.1	10.0	40	169
191	*average, raw, weighed with core*	Calculated from 190	0.91	76.3	0.05	0.3	0.1	9.1	36	155
192	*-, raw, peeled*	Literature sources and calculation from 190; flesh only	1.00	83.8	0.05	0.3	0.1	10.4	41	175
193	*-, raw, peeled, weighed with skin and core*	Calculated from 192	0.70	58.7	0.03	0.2	0.1	7.3	29	124
194	*-, stewed with sugar*	Calculation from 1000g fruit, 190g water, 120g sugar	1.00	74.3	0.04	0.2	0.1	21.9	83	356
195	*-, stewed without sugar*	Calculated from 1000g fruit, 190g water	1.00	85.8	0.04	0.3	0.1	9.1	35	152
196	*dried*	3 samples, 3 brands and calculation from 192	1.00	18.4	0.25	1.6	0.5	52.4	207	884
197	*canned in juice*	10 samples, 7 brands	1.00	86.8	0.04	0.3	Tr	8.5	33	141
198	*canned in syrup*	10 samples, 8 brands	1.00	82.6	0.04	0.2	Tr	13.2	50	215
199	*Comice, raw*	10 samples; flesh and skin	1.00	84.1	0.05	0.3	Tr	8.5	33	141
200	*raw, weighed with core*	Calculated from 199	0.91	76.5	0.05	0.3	Tr	7.7	30	128
201	*Conference, raw*	10 samples; flesh and skin	1.00	83.2	0.05	0.3	0.3	13.2	53	227
202	*raw, weighed with core*	Calculated from 201	0.91	75.7	0.05	0.3	0.3	12.0	49	208
203	*William, raw*	10 samples; flesh and skin	1.00	84.1	0.06	0.4	0.1	8.3	34	143
204	*raw, weighed with core*	Calculated from 203	0.90	75.7	0.05	0.3	0.1	7.5	30	129
205	**Nashi,** *raw*	6 samples; flesh and skin	1.00	87.1	0.05	0.3	0.1	7.1	29	122
206	*weighed with core*	Calculated from 205	0.89	77.5	0.04	0.3	0.1	6.3	26	110
207	**Phyalis**	Ref. 4; fruit only	0.87	82.9	0.29	1.8	0.2	11.1	53	218

No. 14-	Food	Starch	Total sugars	Individual sugars					Dietary fibre		Fibre fractions			
				Gluc	Fruct	Sucr	Malt	Lact	Southgate method	Englyst method	Cellulose	Non-cellulosic polysaccharide		Lignin
												Soluble	Insoluble	
190	**Pears**, average, raw	0	10.0	2.3	7.1	0.7	0	0	N	2.2	0.7	0.7	0.8	N
191	average, raw, weighed with core	0	9.1	2.1	6.5	0.6	0	0	N	2.0	0.6	0.6	0.7	N
192	-, raw, peeled	0	10.4	2.4	7.4	0.7	0	0	2.1	1.7	0.5	0.6	0.5	N
193	-, raw, peeled, weighed with skin and core	0	7.3	1.7	5.2	0.5	0	0	1.5	1.1	0.3	0.4	0.3	N
194	-, stewed with sugar	0	21.9	2.6	6.4	13.0	0	0	1.6	1.3	0.4	0.5	0.4	N
195	-, stewed without sugar	0	9.1	2.1	6.5	0.5	0	0	1.8	1.5	0.4	0.5	0.4	N
196	dried	0	52.4	11.9	36.8	3.6	0	0	10.6	8.3	2.6	3.0	2.6	Tr
197	canned in juice	0	8.5	2.3	5.7	0.6	0	0	(1.5)	1.4	0.6	0.4	0.4	Tr
198	canned in syrup	0	13.2	3.4	6.1	3.4	0.2	0	1.5	1.1	0.5	0.3	0.3	N
199	Comice, raw	0	8.5	1.0	7.1	0.4	0	0	N	2.0	0.7	0.6	0.7	N
200	raw, weighed with core	0	7.7	0.9	6.5	0.4	0	0	N	1.8	0.6	0.5	0.6	N
201	Conference, raw	0	13.2	4.4	7.3	1.5	0	0	N	2.4	0.7	0.8	0.9	N
202	raw, weighed with core	0	12.0	4.0	6.6	1.4	0	0	N	2.2	0.6	0.7	0.8	N
203	William, raw	0	8.3	1.4	6.8	0.1	0	0	N	(2.2)	(0.7)	(0.7)	(0.8)	N
204	raw, weighed with core	0	7.5	1.3	6.1	0.1	0	0	N	(2.0)	(0.6)	(0.6)	(0.7)	N
205	**Nashi**, raw	0	7.1	2.2	4.9	Tr	0	0	N	1.5	0.5	0.5	0.6	N
206	weighed with core	0	6.3	2.0	4.4	Tr	0	0	N	1.3	0.4	0.4	0.5	N
207	**Phyalis**	Tr	*11.1*	N	N	N	0	0	N	N	N	N	N	N

Fruit continued

Inorganic constituents per 100g

No. 14-	Food	mg											μg	
		Na	K	Ca	Mg	P	Fe	Cu	Zn	S	Cl	Mn	Se	I
190	**Pears,** *average, raw*	3	150	11	7	13	0.2	0.06	0.1	5	1	Tr	Tr	1
191	*average, raw, weighed with core*	3	140	10	6	12	0.2	0.05	0.1	5	1	Tr	Tr	1
192	*-, raw, peeled*	3	150	11	7	13	0.2	0.06	0.1	5	1	Tr	Tr	1
193	*-, raw, peeled, weighed with skin and core*	2	110	8	5	9	0.1	0.04	0.1	3	1	Tr	Tr	1
194	*-, stewed with sugar*	2	110	8	5	10	0.1	0.05	0.1	3	Tr	Tr	Tr	1
195	*-, stewed without sugar*	2	130	9	6	11	0.2	0.05	0.1	4	Tr	Tr	Tr	1
196	*dried*	15	750	55	35	65	1.0	0.30	0.5	N	5	0.2	(2)	5
197	*canned in juice*	3	81	6	5	10	0.2	Tr	0.1	(1)	(3)	Tr	Tr	Tr
198	*canned in syrup*	3	68	6	4	7	0.2	0.02	0.1	1	3	Tr	Tr	Tr
199	*Comice, raw*	3	150	12	7	13	0.2	0.04	0.1	5	Tr	Tr	Tr	1
200	*raw, weighed with core*	3	140	11	6	12	0.2	0.04	0.1	5	Tr	Tr	Tr	1
201	*Conference, raw*	4	150	11	7	13	0.2	0.06	0.1	5	2	Tr	Tr	1
202	*raw, weighed with core*	4	140	10	6	12	0.2	0.05	0.1	5	2	Tr	Tr	1
203	*William, raw*	2	150	9	8	12	0.1	0.08	0.2	5	1	0.1	Tr	1
204	*raw, weighed with core*	2	130	8	7	11	0.1	0.07	0.2	5	1	0.1	Tr	1
205	**Nashi,** *raw*	6	110	5	7	11	0.1	0.08	0.1	(5)	1	0.1	N	(1)
206	*weighed with core*	5	98	4	6	10	0.1	0.07	0.1	(4)	1	0.1	N	(1)
207	**Phyalis**	1	320	10	31	67	2.0	0.19	N	43	12	N	N	N

Fruit *continued*

No. 14-	Food	Retinol µg	Carotene µg	Vitamin D µg	Vitamin E mg	Thiamin mg	Ribo-flavin mg	Niacin mg	Trypt 60 mg	Vitamin B6 mg	Vitamin B12 µg	Folate µg	Panto-thenate mg	Biotin µg	Vitamin C mg
190	**Pears,** *average, raw*	0	18	0	0.50	0.02	0.03	0.2	Tr	0.02	0	2	0.07	0.2	6
191	*average, raw, weighed with core*	0	16	0	0.45	0.02	0.03	0.2	Tr	0.02	0	2	0.06	0.2	5
192	*-, raw, peeled*	0	19	0	Tr	0.02	0.03	0.2	Tr	0.02	0	2	0.07	0.2	6
193	*-, raw, peeled, weighed with skin and core*	0	14	0	Tr	0.01	0.02	0.1	Tr	0.01	0	1	0.05	0.2	4
194	*-, stewed with sugar*	0	14	0	Tr	0.01	0.02	0.1	Tr	0.01	0	Tr	0.04	0.1	3
195	*-, stewed without sugar*	0	16	0	Tr	0.01	0.02	0.1	Tr	0.01	0	Tr	0.05	0.1	4
196	*dried*	0	91	0	Tr	N	N	N	Tr	N	0	Tr	N	N	Tr
197	*canned in juice*	0	Tr	0	Tr	0.01	0.01	0.2	Tr	0.03	0	4	0.04	0.2	3
198	*canned in syrup*	0	Tr	0	Tr	0.01	0.01	0.2	Tr	0.03	0	3	0.04	0.2	2
199	*Comice, raw*	0	15	0	0.50	0.02	0.03	0.2	Tr	0.02	0	2	0.07	0.2	6
200	*raw, weighed with core*	0	14	0	0.45	0.02	0.03	0.2	Tr	0.02	0	2	0.06	0.2	5
201	*Conference, raw*	0	15	0	0.50	0.02	0.03	0.2	Tr	0.02	0	2	0.07	0.2	6
202	*raw, weighed with core*	0	14	0	0.45	0.02	0.03	0.2	Tr	0.02	0	2	0.06	0.2	5
203	*William, raw*	0	25	0	0.50	0.02	0.03	0.2	Tr	0.02	0	2	0.07	0.2	6
204	*raw, weighed with core*	0	23	0	0.45	0.02	0.03	0.2	Tr	0.02	0	2	0.06	0.2	5
205	**Nashi,** *raw*	0	Tr	0	(0.50)	0.03	0.03	0.3	Tr	0.04	0	3	(0.07)	(0.2)	4
206	*weighed with core*	0	Tr	0	(0.45)	0.03	0.03	0.3	Tr	0.04	0	3	(0.06)	(0.2)	4
207	**Phyalis**	0	(1430)	0	N	0.05	0.02	0.3	0.3	N	0	N	N	N	49

61

No. 14-	Food	Description and main data sources	Edible proportion	Water g	Total nitrogen g	Protein g	Fat g	Carbohydrate g	Energy value kcal	kJ
208	**Pineapple**, *raw*	10 samples; flesh only	1.00	86.5	0.06	0.4	0.2	10.1	41	176
209	*raw, weighed with skin and top*	Calculated from 208	0.53	45.8	0.03	0.2	0.1	5.3	22	92
210	*dried*	2 samples and calculation from 208	1.00	9.3	0.41	2.5	1.3	67.9	276	1177
211	*canned in juice*	10 samples, 10 brands; cubes and slices	1.00	86.8	0.05	0.3	Tr	12.2	47	200
212	*canned in syrup*	10 samples, 10 brands; cubes and slices	1.00	82.2	0.08	0.5	Tr	16.5	64	273
213	**Plums**, *average, raw*	Assorted varieties; flesh and skin	1.00	83.9	0.09	0.6	0.1	8.8	36	155
214	*average, raw, weighed with stones*	Calculated from 213	0.94	78.9	0.08	0.5	0.1	8.3	34	145
215	*-, stewed with sugar*	1350g fruit, 100g water, 162g sugar; stones removed	1.00	74.2	0.08	0.5	0.1	20.2	79	335
216	*-, stewed with sugar, weighed with stones*	Calculated from 215	0.95	70.5	0.08	0.5	0.1	19.2	75	319
217	*-, stewed without sugar*	500g fruit, 100g water and calculation from 215; stones removed	1.00	85.3	0.07	0.5	0.1	7.3	30	129
218	*-, stewed without sugar, weighed with stones*	Calculated from 217	0.95	81.0	0.07	0.4	0.1	6.9	29	121
219	*canned in syrup*	10 samples, 7 brands; red, golden and Victoria; no stones	1.00	81.4	0.05	0.3	Tr	15.5	59	253
220	*Victoria, raw*	Flesh and skin	1.00	86.0	0.09	0.6	0.1	9.6	39	168
221	*raw, weighed with stones*	Calculated from 220	0.94	80.8	0.08	0.5	0.1	9.0	37	156
222	*stewed without sugar*	700g fruit, 100g water; stones removed	1.00	87.2	0.08	0.5	0.1	8.3	34	145
223	*stewed without sugar, weighed with stones*	Calculated from 222	0.95	82.8	0.08	0.5	0.1	7.9	33	139
224	*yellow, raw*	Ref. 2; flesh and skin	1.00	89.0	0.08	0.5	0.1	5.9	25	107
225	*raw, weighed with stones*	Calculated from 224	0.96	85.4	0.08	0.5	0.1	5.7	24	103

Fruit continued

Carbohydrate fractions, g per 100g

No. Food 14-	Starch	Total sugars	Individual sugars					Dietary fibre		Fibre fractions			
			Gluc	Fruct	Sucr	Malt	Lact	Southgate method	Englyst method	Cellulose	Non-cellulosic polysaccharide Soluble	Insoluble	Lignin
208 **Pineapple**, raw	0	10.1	2.0	2.5	5.5	0	0	1.3	1.2	0.5	0.1	0.6	0.1
209 raw, weighed with skin and top	0	5.3	1.1	1.3	2.9	0	0	0.7	0.6	0.3	0.1	0.3	0.1
210 dried	0	67.9	13.6	17.0	37.3	0	0	8.8	8.1	3.4	0.7	4.1	0.7
211 canned in juice	0	12.2	4.0	4.0	4.2	0	0	(0.8)	0.5	0.2	0.1	0.2	Tr
212 canned in syrup	0	16.5	6.0	4.8	5.8	0	0	0.8	0.7	0.3	0.1	0.3	Tr
213 **Plums**, average, raw	0	8.8	4.3	2.0	2.5	0	0	2.3	1.6	0.3	1.1	0.2	0.1
214 average, raw, weighed with stones	0	8.3	4.0	1.9	2.3	0	0	2.2	1.5	0.3	1.0	0.2	0.1
215 -, stewed with sugar	0	20.2	4.5	2.5	13.3	0	0	1.9	1.3	0.3	0.9	0.2	0.1
216 -, stewed with sugar, weighed with stones	0	19.2	4.3	2.4	12.6	0	0	1.8	1.2	0.3	0.8	0.2	0.1
217 -, stewed without sugar	0	7.3	3.6	1.8	1.9	0	0	1.9	1.3	0.3	0.9	0.2	0.1
218 -, stewed without sugar, weighed with stones	0	6.9	3.4	1.7	1.8	0	0	1.8	1.2	0.3	0.8	0.2	0.1
219 canned in syrup	0	15.5	7.1	6.2	2.2	0	0	1.0	0.8	0.1	0.5	0.1	Tr
220 Victoria, raw	0	9.6	4.7	2.2	2.7	0	0	1.9	1.8	0.3	1.2	0.3	0.1
221 raw, weighed with stones	0	9.0	4.4	2.1	2.5	0	0	1.8	1.7	0.3	1.1	0.3	0.1
222 stewed without sugar	0	8.3	4.2	2.0	2.1	0	0	1.7	1.6	0.3	1.0	0.3	0.1
223 stewed without sugar, weighed with stones	0	7.9	4.0	1.9	2.0	0	0	1.6	1.5	0.3	0.9	0.3	0.1
224 yellow, raw	0	5.9	2.1	1.4	2.4	0	0	N	1.0	0.3	0.6	0.1	Tr
225 raw, weighed with stones	0	5.7	2.0	1.3	2.3	0	0	N	1.0	0.3	0.6	0.1	Tr

Fruit continued

Inorganic constituents per 100g

No. 14-	Food	Na	K	Ca	Mg	P	Fe	Cu	Zn	S	Cl	Mn	Se	I
							mg						µg	
		Na	K	Ca	Mg	P	Fe	Cu	Zn	S	Cl	Mn	Se	I
208	**Pineapple**, raw	2	160	18	16	10	0.2	0.11	0.1	3	29	0.5	Tr	Tr
209	raw, weighed with skin and top	1	85	9	8	5	0.1	0.06	0.1	2	15	0.3	Tr	Tr
210	dried	13	1080	120	110	67	1.3	0.74	0.7	N	200	3.4	Tr	Tr
211	canned in juice	1	71	8	13	5	0.5	0.08	0.1	(3)	(4)	0.9	Tr	Tr
212	canned in syrup	2	79	6	11	5	0.2	0.02	0.1	3	4	0.9	Tr	Tr
213	**Plums**, average, raw	2	240	13	8	23	0.4	0.10	0.1	5	Tr	0.1	Tr	Tr
214	average, raw, weighed with stones	2	230	12	7	22	0.4	0.09	0.1	5	Tr	0.1	Tr	Tr
215	-, stewed with sugar	2	200	11	7	19	0.3	0.08	0.1	4	Tr	0.1	Tr	Tr
216	-, stewed with sugar, weighed with stones	2	190	10	7	18	0.3	0.08	0.1	4	Tr	0.1	Tr	Tr
217	-, stewed without sugar	2	200	11	7	19	0.3	0.08	0.1	4	Tr	0.1	Tr	Tr
218	-, stewed without sugar, weighed with stones	2	190	10	7	18	0.3	0.08	0.1	4	Tr	0.1	Tr	Tr
219	canned in syrup	6	79	9	4	10	N	Tr	Tr	N	N	Tr	Tr	Tr
220	Victoria, raw	2	190	11	7	16	0.4	0.10	Tr	4	Tr	0.1	Tr	Tr
221	raw, weighed with stones	2	180	10	7	15	0.4	0.09	Tr	4	Tr	0.1	Tr	Tr
222	stewed without sugar	2	170	10	6	14	0.3	0.09	Tr	3	Tr	0.1	Tr	Tr
223	stewed without sugar, weighed with stones	2	160	9	6	13	0.3	0.09	Tr	3	Tr	0.1	Tr	Tr
224	yellow, raw	1	190	7	5	(23)	0.2	(0.10)	0.1	(5)	Tr	(0.1)	Tr	Tr
225	raw, weighed with stones	1	180	7	5	(22)	0.2	(0.10)	0.1	(5)	Tr	(0.1)	Tr	Tr

No. 14-	Food	Retinol μg	Carotene μg	Vitamin D μg	Vitamin E mg	Thiamin mg	Ribo-flavin mg	Niacin mg	Trypt 60 mg	Vitamin B6 mg	Vitamin B12 μg	Folate μg	Panto-thenate mg	Biotin μg	Vitamin C mg
208	**Pineapple**, raw	0	18	0	0.10	0.08	0.03	0.3	0.1	0.09	0	5	0.16	0.3	12
209	raw, weighed with skin and top	0	9	0	0.05	0.04	0.02	0.2	0.1	0.05	0	3	0.08	0.2	6
210	dried	0	(120)	0	0.68	N	N	N	0.5	N	0	Tr	N	N	Tr
211	canned in juice	0	12	0	(0.05)	0.09	0.01	0.2	0.1	0.09	0	1	0.11	0.1	11
212	canned in syrup	0	11	0	0.06	0.07	0.01	0.2	0.1	0.07	0	(1)	0.07	0.1	13
213	**Plums**, average, raw	0	295	0	0.61	0.05	0.03	1.1	0.1	0.05	0	3	0.15	Tr	4
214	average, raw, weighed with stones	0	275	0	0.57	0.05	0.03	1.0	0.1	0.05	0	3	0.14	Tr	4
215	-, stewed with sugar	0	65	0	0.51	0.03	0.02	0.7	0.1	0.03	0	Tr	0.09	Tr	3
216	-, stewed with sugar, weighed with stones	0	62	0	0.48	0.03	0.02	0.7	0.1	0.03	0	Tr	0.09	Tr	3
217	-, stewed without sugar	0	65	0	0.50	0.03	0.02	0.7	0.1	0.03	0	Tr	0.09	Tr	3
218	-, stewed without sugar, weighed with stones	0	62	0	0.47	0.03	0.02	0.7	0.1	0.03	0	Tr	0.09	Tr	3
219	canned in syrup	0	29	0	0.25	0.01	0.01	0.3	Tr	(0.02)	0	Tr	(0.04)	Tr	1
220	Victoria, raw	0	(295)	0	0.61	0.05	0.03	0.5	0.1	0.05	0	3	0.15	Tr	6
221	raw, weighed with stones	0	(275)	0	0.57	0.05	0.03	0.5	0.1	0.05	0	3	0.14	Tr	6
222	stewed without sugar	0	69	0	0.53	0.03	0.02	0.3	0.1	0.03	0	2	0.10	Tr	4
223	stewed without sugar, weighed with stones	0	65	0	0.50	0.03	0.02	0.3	0.1	0.03	0	2	0.10	Tr	4
224	yellow, raw	0	125	0	(0.61)	0.03	0.05	0.6	0.1	(0.05)	0	(3)	(0.15)	Tr	5
225	raw, weighed with stones	0	120	0	(0.59)	0.03	0.05	0.6	0.1	(0.05)	0	(3)	(0.14)	Tr	5

No. 14-	Food	Description and main data sources	Edible proportion	Water g	Total nitrogen g	Protein g	Fat g	Carbo-hydrate g	Energy value kcal	kJ
226	**Pomegranate**	Analysis and literature sources; flesh and pips	1.00	80.0	0.21	1.3	0.2	11.8	51	218
227	*weighed with skin*	Calculated from 226	0.65	52.0	0.14	0.9	0.1	7.7	33	142
228	**Pomelo**	Literature sources; flesh only	1.00	90.2	0.10	0.6	0.2	(6.8)	30	126
229	*weighed with peel and pips*	Calculated from 228	0.61	55.0	0.06	0.4	0.1	(4.1)	18	76
230	**Prickly pears**	Literature sources; flesh and seeds	0.61	85.2	0.11	0.7	0.3	11.5	49	207
231	**Prunes**	No stones	1.00	22.1	0.45	2.8	0.5	38.4	160	681
232	*weighed with stones*	Calculated from 231	0.84	18.6	0.38	2.4	0.4	32.3	134	572
233	*stewed with sugar*	Calculated from 450g fruit, 450g water, 54g sugar	1.00	56.0	0.21	1.3	0.2	25.5	103	439
234	*stewed with sugar, weighed with stones*	Calculated from 233	0.92	51.5	0.19	1.2	0.2	23.5	95	405
235	*stewed without sugar*	Calculated from 450g fruit, 450g water	1.00	60.4	0.23	1.4	0.3	19.5	81	346
236	*stewed without sugar, weighed with stones*	Calculated from 235	0.91	55.0	0.21	1.3	0.2	17.8	74	314
237	*canned in juice*	10 samples; stones removed	0.93	74.1	0.12	0.7	0.2	19.7	79	335
238	*canned in syrup*	11 samples, 6 brands; stones removed	0.92	69.9	0.10	0.6	0.2	23.0	90	386
239	*ready-to-eat*	4 samples; semi-dried	1.00	31.1	0.40	2.5	0.4	34.0	141	601
240	*weighed with stones*	Calculated from 239	0.86	26.7	0.34	2.1	0.3	29.2	121	514
241	**Quinces**	Flesh only	0.69	84.2	0.05	0.3	0.1	6.3	26	110

Fruit *continued*

Carbohydrate fractions, g per 100g

No. 14-	Food	Starch	Total sugars	Individual sugars					Dietary fibre		Fibre fractions			
				Gluc	Fruct	Sucr	Malt	Lact	Southgate method	Englyst method	Cellulose	Non-cellulosic polysaccharide		Lignin
												Soluble	Insoluble	
226	**Pomegranate**	0	11.8	6.4	5.2	0.2	0	0	N	3.4	1.5	0.7	1.2	N
227	*weighed with skin*	0	7.7	4.2	3.4	0.1	0	0	N	2.2	1.0	0.5	0.8	N
228	**Pomelo**	0	(6.8)	(2.1)	(2.3)	(2.4)	0	0	0.8	N	0.2	N	N	0.1
229	*weighed with peel and pips*	0	(4.1)	(1.3)	(1.4)	(1.5)	0	0	0.5	N	0.1	N	N	0.1
230	**Prickly pears**	0	11.5	6.0	5.4	0.1	0	0	N	N	N	N	N	N
231	**Prunes**	0	38.4	20.2	13.7	4.5	0	0	14.5	6.5	1.1	4.4	0.9	2.0
232	*weighed with stones*	0	32.3	16.7	11.8	3.9	0	0	12.2	5.5	0.9	3.7	0.8	1.7
233	*stewed with sugar*	0	25.5	10.1	7.0	8.4	0	0	6.9	3.1	0.5	2.1	0.4	1.0
234	*stewed with sugar, weighed with stones*	0	23.5	9.3	6.4	7.7	0	0	6.3	2.9	0.5	1.9	0.4	0.9
235	*stewed without sugar*	0	19.5	10.4	7.1	2.1	0	0	7.4	3.3	0.6	2.2	0.5	1.0
236	*stewed without sugar, weighed with stones*	0	17.8	9.4	6.4	1.9	0	0	6.7	3.0	0.5	2.0	0.5	0.9
237	*canned in juice*	0	19.7	10.2	8.4	1.1	0	0	N	2.4	0.4	1.5	0.5	Tr
238	*canned in syrup*	0	23.0	11.0	5.5	6.5	0	0	N	2.8	0.5	1.7	0.6	Tr
239	*ready-to-eat*	0	34.0	17.9	12.1	4.1	0	0	12.8	5.7	1.0	3.9	0.8	1.8
240	*weighed with stones*	0	29.2	15.4	10.4	3.4	0	0	11.0	4.9	0.9	3.3	0.7	1.5
241	**Quinces**	Tr	6.3	2.3	3.7	0.3	0	0	5.8	N	N	N	N	N

No. 14-	Food	Na	K	Ca	Mg	P	Fe	Cu	Zn	S	Cl	Mn	Se	I
							mg						µg	
226	**Pomegranate**	2	240	12	11	29	0.7	0.17	0.4	12	2	N	N	N
227	weighed with skin	1	160	8	7	19	0.5	0.11	0.3	8	1	N	N	N
228	**Pomelo**	1	230	27	6	20	0.3	0.05	0.1	(7)	(3)	Tr	N	N
229	weighed with peel and pips	1	140	16	4	12	0.2	0.03	0.1	(4)	(2)	Tr	N	N
230	**Prickly pears**	4	210	53	57	27	0.4	N	0.6	N	N	N	N	N
231	**Prunes**	12	860	38	27	83	2.9	0.16	0.5	19	3	0.3	3	N
232	weighed with stones	10	720	32	23	70	2.4	0.13	0.4	16	3	0.3	3	N
233	stewed with sugar	5	410	18	12	39	1.4	0.08	0.3	9	1	0.1	1	N
234	stewed with sugar, weighed with stones	5	380	17	11	36	1.3	0.07	0.2	8	1	0.1	1	N
235	stewed without sugar	6	440	19	13	42	1.5	0.08	0.3	9	1	0.1	1	N
236	stewed without sugar, weighed with stones	5	400	17	12	38	1.3	0.07	0.2	8	1	0.1	1	N
237	canned in juice	18	340	26	15	30	2.2	0.09	1.0	N	N	0.1	Tr	N
238	canned in syrup	(18)	(340)	(26)	(15)	(30)	(2.2)	(0.09)	(1.0)	N	N	(0.1)	Tr	N
239	ready-to-eat	11	760	34	24	73	2.6	0.14	0.4	17	3	0.3	3	N
240	weighed with stones	9	650	29	21	63	2.2	0.12	0.3	15	3	0.3	3	N
241	**Quinces**	3	200	14	6	19	0.3	0.13	0.5	5	2	N	N	N

Fruit continued

No. 14-	Food	Retinol µg	Carotene µg	Vitamin D µg	Vitamin E mg	Thiamin mg	Ribo-flavin mg	Niacin mg	Trypt 60 mg	Vitamin B6 mg	Vitamin B12 µg	Folate µg	Panto-thenate mg	Biotin µg	Vitamin C mg
226	**Pomegranate**	0	33	0	N	0.05	0.04	0.3	0.2	0.31	0	N	0.57	N	13
227	*weighed with skin*	0	21	0	N	0.03	0.03	0.2	0.1	0.20	0	N	0.37	N	8
228	**Pomelo**	0	23[a]	0	(0.09)	0.05	0.03	0.2	0.1	0.04	0	(26)	(0.28)	(1.0)	45
229	*weighed with peel and pips*	0	14	0	(0.05)	0.03	0.02	0.1	0.1	0.02	0	(16)	(0.17)	(0.6)	27
230	**Prickly pears**	0	45	0	N	0.02	0.04	0.4	N	N	0	N	N	N	22
231	**Prunes**	0	155	0	N	0.10	0.20	1.5	0.5	0.24	0	4	0.46	Tr	Tr
232	*weighed with stones*	0	130	0	N	0.08	0.17	1.3	0.4	0.20	0	3	0.39	Tr	Tr
233	*stewed with sugar*	0	73	0	N	0.04	0.07	0.5	0.2	0.09	0	Tr	0.16	Tr	Tr
234	*stewed with sugar, weighed with stones*	0	67	0	N	0.04	0.06	0.5	0.2	0.08	0	Tr	0.15	Tr	Tr
235	*stewed without sugar*	0	78	0	N	0.04	0.08	0.6	0.3	0.10	0	Tr	0.18	Tr	Tr
236	*stewed without sugar, weighed with stones*	0	72	0	N	0.04	0.07	0.5	0.2	0.09	0	Tr	0.16	Tr	Tr
237	*canned in juice*	0	140	0	N	0.02	0.02	0.5	0.1	(0.06)	0	5	(0.07)	Tr	Tr
238	*canned in syrup*	0	(140)	0	N	(0.02)	(0.02)	(0.5)	0.1	(0.05)	0	(5)	(0.06)	Tr	Tr
239	*ready-to-eat*	0	140	0	N	0.09	0.18	1.3	0.4	0.21	0	3	0.41	Tr	Tr
240	*weighed with stones*	0	120	0	N	0.08	0.15	1.1	0.3	0.18	0	3	0.35	Tr	Tr
241	**Quinces**	0	Tr	0	N	0.02	0.02	0.2	Tr	0.04	0	N	0.08	(0.1)	15

a Pink varieties contain approximately 380µg carotene per 100g

Fruit *continued*

Composition of food per 100g

No. 14-	Food	Description and main data sources	Edible proportion	Water g	Total nitrogen g	Protein g	Fat g	Carbo-hydrate g	Energy value kcal	kJ
242	**Raisins**	10 samples, 8 brands. Large stoned variety	1.00	13.2	0.34	2.1	0.4	69.3	272	1159
243	**Rambutan**	Literature sources; flesh only	0.40	80.4	0.16	1.0	0.4	16.3	69	293
244	**Raspberries,** *raw*	9 samples; whole fruit	1.00	87.0	0.22	1.4	0.3	4.6	25	109
245	*stewed with sugar*	Calculated from 700g fruit, 105g water, 84g sugar	1.00	78.1	0.19	1.2	0.3	15.0	63	271
246	*stewed without sugar*	Calculated from 700g fruit, 105g water	1.00	87.4	0.21	1.4	0.3	4.4	24	105
247	*frozen*	10 samples, 5 brands	1.00	86.2	0.19	1.2	0.3	4.9	26	110
248	*canned in syrup*	Mixed sample	1.00	74.0	0.10	0.6	0.1	22.5	88	374
249	**Redcurrants,** *raw*	Whole fruit, stalks removed	0.97	82.8	0.18	1.1	Tr	4.4	21	89
250	*stewed with sugar*	Calculated from 700g fruit, 210g water, 84g sugar	1.00	77.0	0.14	0.9	Tr	13.3	53	227
251	*stewed without sugar*	Calculated from 700g fruit, 210g water	1.00	85.1	0.15	0.9	Tr	3.8	17	76
252	**Rhubarb,** *raw*	Stems only	0.87	94.2	0.14	0.9	0.1	0.8	7	32
253	*stewed with sugar*	1000g fruit, 100g water, 120g sugar	1.00	84.6	0.14	0.9	0.1	11.5	48	203
254	*stewed without sugar*	500g fruit, 50g water and calculation from 253	1.00	94.1	0.15	0.9	0.1	0.7	7	30
255	*canned in syrup*	10 samples, 6 brands	1.00	90.6	0.08	0.5	Tr	7.6	31	130
256	**Sapodilla**	Literature sources; flesh only	0.81	76.3	0.10	0.6	0.9	15.5	69	291
257	**Satsumas**	10 samples; flesh only	1.00	87.4	0.14	0.9	0.1	8.5	36	155
258	*weighed with peel*	Calculated from 257	0.71	62.1	0.10	0.6	0.1	6.0	26	110

Fruit *continued*

Carbohydrate fractions, g per 100g

No. 14-	Food	Starch	Total sugars	Individual sugars					Dietary fibre		Cellulose	Fibre fractions Non-cellulosic polysaccharide		Lignin
				Gluc	Fruct	Sucr	Malt	Lact	Southgate method	Englyst method		Soluble	Insoluble	
242	**Raisins**	0	69.3	34.5	34.8	Tr	0	0	6.1	2.0	0.8	1.0	0.2	N
243	**Rambutan**	0	16.3	2.9	3.1	10.3	0	0	1.3	N	0.3	N	N	Tr
244	**Raspberries**, *raw*	0	4.6	1.9	2.4	0.2	0	0	6.7	2.5	1.2	0.7	0.6	2.2
245	*stewed with sugar*	0	15.0	2.2	2.7	10.1	0	0	5.9	2.2	1.0	0.6	0.5	1.9
246	*stewed without sugar*	0	4.4	1.8	2.3	0.2	0	0	6.5	2.4	1.2	0.7	0.6	2.1
247	*frozen*	0	4.9	2.0	2.6	0.2	0	0	7.1	2.7	1.3	0.7	0.6	2.3
248	*canned in syrup*	0	22.5	N	N	N	0	0	(4.5)	1.5	0.6	0.3	0.6	0.9
249	**Redcurrants**, *raw*	0	4.4	1.7	2.6	0.1	0	0	7.4	3.4	0.6	0.7	2.1	1.3
250	*stewed with sugar*	0	13.3	1.8	2.5	8.9	0	0	5.8	2.7	0.5	0.5	1.6	1.0
251	*stewed without sugar*	0	3.8	1.5	2.2	0.1	0	0	6.3	2.9	0.5	0.6	1.8	1.1
252	**Rhubarb**, *raw*	0	0.8	0.4	0.4	0.1	0	0	2.3	1.4	0.8	0.5	0.1	0.1
253	*stewed with sugar*	0	11.5	1.2	1.2	9.1	0	0	2.0	1.2	0.6	0.5	0.1	0.1
254	*stewed without sugar*	0	0.7	0.3	0.3	0.1	0	0	2.1	1.3	0.7	0.5	0.1	0.1
255	*canned in syrup*	0	7.6	2.9	2.6	2.1	0	0	1.3	0.8	0.4	0.3	0.1	0.1
256	**Sapodilla**	0.8	14.7	6.7	5.3	2.7	0	0	8.2	N	N	N	N	N
257	**Satsumas**	0	8.5	1.5	1.8	5.1	0	0	(1.7)	1.3	0.3	0.9	0.1	Tr
258	*weighed with peel*	0	6.0	1.1	1.3	3.6	0	0	(1.2)	0.9	0.2	0.6	0.1	Tr

Fruit continued

Inorganic constituents per 100g

No. 14-	Food	Na	K	Ca	Mg	P	Fe	Cu	Zn	S	Cl	Mn	Se	I
							mg						µg	
242	**Raisins**	60	1020	46	35	76	3.8	0.39	0.7	23	9	0.3	(8)	N
243	**Rambutan**	1	100	14	10	15	0.1	N	0.6	N	N	N	N	N
244	**Raspberries**, raw	3	170	25	19	31	0.7	0.10	0.3	17	22	0.4	N	N
245	stewed with sugar	2	150	22	16	27	0.6	0.09	0.3	14	19	0.3	N	N
246	stewed without sugar	2	160	24	18	29	0.7	0.10	0.3	16	21	0.4	N	N
247	frozen	12	160	28	22	37	0.8	0.06	0.4	17	22	0.5	N	N
248	canned in syrup	4	100	14	11	14	1.7	0.10	N	N	5	0.3	N	N
249	**Redcurrants**, raw	2	280	36	13	30	1.2	0.12	0.2	29	14	0.2	N	N
250	stewed with sugar	1	220	28	10	23	0.9	0.10	0.2	22	10	0.2	N	N
251	stewed without sugar	1	240	30	11	25	1.0	0.10	0.2	24	11	0.2	N	N
252	**Rhubarb**, raw	3	290	93	13	17	0.3	0.07	0.1	8	87	0.2	Tr	N
253	stewed with sugar	1	210	33	6	18	0.1	0.02	Tr	7	75	0.3	Tr	N
254	stewed without sugar	1	230	35	6	19	0.1	0.02	Tr	7	79	0.3	Tr	N
255	canned in syrup	4	89	36	5	8	0.8	Tr	0.1	N	15	0.1	Tr	N
256	**Sapodilla**	7	190	27	26	15	1.1	0.36	N	17	26	N	N	N
257	**Satsumas**	4	130	31	10	18	0.1	0.01	0.1	(10)	(2)	Tr	N	N
258	weighed with peel	3	92	22	7	13	0.1	0.01	0.1	(7)	(1)	Tr	N	N

Fruit *continued*

No. 14-	Food	Retinol µg	Carotene µg	Vitamin D µg	Vitamin E mg	Thiamin mg	Riboflavin mg	Niacin mg	Trypt 60 mg	Vitamin B6 mg	Vitamin B12 µg	Folate µg	Pantothenate mg	Biotin µg	Vitamin C mg
242	**Raisins**	0	12	0	N	0.12	0.05	0.6	0.2	0.25	0	10	0.15	2.0	1
243	**Rambutan**	0	0	0	N	0.02	0.06	0.6	0.1	N	0	N	N	N	78
244	**Raspberries**, *raw*	0	6	0	0.48	0.03	0.05	0.5	0.3	0.06	0	33	0.24	1.9	32
245	*stewed with sugar*	0	5	0	0.42	0.02	0.03	0.3	0.3	0.04	0	5	0.16	1.2	21
246	*stewed without sugar*	0	5	0	0.46	0.02	0.04	0.4	0.3	0.05	0	6	0.17	1.4	23
247	*frozen*	0	4	0	0.48	(0.03)	(0.05)	(0.5)	0.2	(0.06)	0	(33)	(0.24)	(1.9)	22
248	*canned in syrup*	0	3	0	0.15	0.01	0.03	0.3	0.1	0.04	0	(10)	0.17	(0.7)	7
249	**Redcurrants**, *raw*	0	25	0	0.10	0.04	(0.06)	0.1	0.2	0.05	0	N	0.06	2.6	40
250	*stewed with sugar*	0	19	0	0.08	0.02	(0.04)	0.1	0.2	0.03	0	N	0.04	1.5	23
251	*stewed without sugar*	0	21	0	0.09	0.03	(0.04)	0.1	0.2	0.03	0	N	0.04	1.7	26
252	**Rhubarb**, *raw*	0	60	0	0.20	0.03	0.03	0.3	0.1	0.02	0	7	0.09	N	6
253	*stewed with sugar*	0	28	0	0.17	0.03	0.02	0.2	0.1	0.02	0	4	0.08	N	5
254	*stewed without sugar*	0	30	0	0.18	0.03	0.02	0.2	0.1	0.02	0	4	0.08	N	5
255	*canned in syrup*	0	(18)	0	(0.11)	0.02	0.01	0.1	0.1	0.01	0	3	0.05	N	3
256	**Sapodilla**	0	(53)	0	N	0.01	0.02	0.2	N	0.04	0	N	0.24	N	10
257	**Satsumas**	0	75	0	N	0.09	0.04	0.3	0.1	(0.07)	0	33	(0.20)	N	27
258	*weighed with peel*	0	53	0	N	0.06	0.03	0.2	0.1	(0.05)	0	23	(0.14)	N	19

No. 14-	Food	Description and main data sources	Edible proportion	Water g	Total nitrogen g	Protein g	Fat g	Carbohydrate g	Energy value kcal	kJ
259	**Sharon fruit**	7 samples; flesh only	0.93	79.9	0.13	0.8	Tr	18.6	73	311
260	**Strawberries**, *raw*	9 samples; flesh and pips	0.95	89.5	0.13	0.8	0.1	6.0	27	113
261	*frozen*	6 samples, 3 brands	1.00	87.7	0.12	0.7	0.1	7.8	33	140
262	*canned in syrup*	10 samples	1.00	81.7	0.07	0.5	Tr	16.9	65	279
263	**Sultanas**	10 samples, 9 brands; whole fruit	1.00	15.2	0.43	2.7	0.4	69.4	275	1171
264	**Tamarillos**	Literature sources; flesh and seeds	0.93	86.5	0.32	2.0	0.3	4.7	28	120
265	**Tamarind**	Literature sources	0.41	35.8	0.37	2.3	0.3	56.5	238	1011
266	**Tangerines**	Flesh only	1.00	86.7	0.14	0.9	0.1	8.0	35	147
267	*weighed with peel and pips*	Calculated from 266	0.73	63.3	0.10	0.7	0.1	5.8	25	108
268	**Whitecurrants**, *raw*	Whole fruit, stalks removed	0.96	83.3	0.20	1.3	Tr	5.6	26	112
269	*stewed with sugar*	Calculated from 700g fruit, 210g water, 84g sugar	1.00	77.5	0.16	1.0	Tr	14.2	57	245
270	*stewed without sugar*	Calculated from 700g fruit, 210g water	1.00	85.7	0.17	1.1	Tr	4.8	22	95

Fruit *continued*

Carbohydrate fractions, g per 100g

No. Food 14-	Starch	Total sugars	Individual sugars					Dietary fibre		Fibre fractions			
			Gluc	Fruct	Sucr	Malt	Lact	Southgate method	Englyst method	Cellulose	Non-cellulosic polysaccharide		Lignin
											Soluble	Insoluble	
259 **Sharon fruit**	0	18.6	9.3	9.3	Tr	0	0	N	1.6	0.5	0.6	0.5	Tr
260 **Strawberries**, *raw*	0	6.0	2.6	3.0	0.3	0	0	2.0	1.1	0.4	0.5	0.2	0.1
261 *frozen*	0	7.8	3.5	4.1	0.2	0	0	(2.0)	1.2	0.4	0.5	0.3	0.1
262 *canned in syrup*	0	16.9	4.7	4.9	7.3	0	0	0.9	0.7	0.3	0.3	0.2	(0.1)
263 **Sultanas**	0	69.4	34.8	34.6	Tr	0	0	6.3	2.0	0.7	0.9	0.3	N
264 **Tamarillos**	Tr	4.7	1.1	1.2	2.4	0	0	N	N	N	N	N	N
265 **Tamarind**	N	N	N	N	N	0	0	N	N	N	N	N	N
266 **Tangerines**	0	8.0	1.4	1.6	5.1	0	0	1.7	1.3	0.3	0.9	0.1	Tr
267 *weighed with peel and pips*	0	5.8	1.0	1.2	3.7	0	0	1.2	0.9	0.2	0.7	0.1	Tr
268 **Whitecurrants**, *raw*	0	5.6	2.4	2.3	0.8	0	0	6.1	(3.4)	(0.6)	(0.7)	(2.1)	(1.3)
269 *stewed with sugar*	0	14.2	2.4	2.3	9.5	0	0	4.8	(2.7)	(0.5)	(0.5)	(1.6)	(1.0)
270 *stewed without sugar*	0	4.8	2.1	2.0	0.6	0	0	5.2	(2.9)	(0.5)	(0.6)	(1.8)	(1.1)

Fruit *continued*

Inorganic constituents per 100g

No. 14-	Food	Na	K	Ca	Mg	P	Fe	Cu	Zn	S	Cl	Mn	Se	I
							mg						µg	
259	**Sharon fruit**	5	210	10	11	19	0.1	0.10	0.1	N	10	0.3	N	N
260	**Strawberries**, *raw*	6	160	16	10	24	0.4	0.07	0.1	13	18	0.3	Tr	9
261	*frozen*	1	170	19	13	25	1.0	0.04	0.1	13	18	0.2	Tr	9
262	*canned in syrup*	9	87	11	7	15	1.1	Tr	0.1	N	(5)	0.2	Tr	N[a]
263	**Sultanas**	19	1060	64	31	86	2.2	0.40	0.3	44	16	0.3	N	N
264	**Tamarillos**	1	300	10	19	43	0.8	0.05	0.1	17	29	0.1	Tr	N
265	**Tamarind**	15	600	77	92	94	1.8	N	N	N	N	N	N	N
266	**Tangerines**	2	160	42	11	17	0.3	0.01	0.1	10	2	Tr	N	N
267	*weighed with peel and pips*	1	120	31	8	12	0.2	0.01	0.1	7	1	Tr	N	N
268	**Whitecurrants**, *raw*	2	290	22	13	28	0.9	0.14	(0.2)	24	11	(0.2)	N	N
269	*stewed with sugar*	1	230	17	10	21	0.7	0.11	(0.2)	17	8	0.2	N	N
270	*stewed without sugar*	1	250	18	11	23	0.8	0.12	(0.2)	20	9	(0.2)	N	N

a Iodine from erythrosine is present but largely unavailable

Fruit *continued*

No. 14-	Food	Retinol µg	Carotene µg	Vitamin D µg	Vitamin E mg	Thiamin mg	Ribo-flavin mg	Niacin mg	Trypt 60 mg	Vitamin B6 mg	Vitamin B12 µg	Folate µg	Panto-thenate mg	Biotin µg	Vitamin C mg
259	**Sharon fruit**	0	950	0	N	0.03	0.05	0.3	N	N	0	7	N	N	19
260	**Strawberries**, *raw*	0	8	0	0.20	0.03	0.03	0.6	0.1	0.06	0	20	0.34	1.1	77
261	*frozen*	0	8	0	0.20	(0.03)	(0.03)	(0.6)	0.1	(0.06)	0	(20)	(0.34)	(1.1)	48
262	*canned in syrup*	0	4	0	N	0.01	0.02	0.3	0.1	0.03	0	6	0.21	(1.0)	29
263	**Sultanas**	0	12	0	0.70	0.09	0.05	0.8	0.2	0.25	0	27	0.09	4.8	Tr
264	**Tamarillos**	0	920	0	1.86	0.06	0.03	0.3	0.3	0.19	0	N	N	N	23
265	**Tamarind**	0	14	0	N	0.29	0.10	1.4	N	0.08	0	N	0.15	N	3
266	**Tangerines**	0	97	0	N	0.07	0.02	0.2	0.1	0.07	0	21	0.20	N	30
267	*weighed with peel and pips*	0	71	0	N	0.05	0.01	0.1	0.1	0.05	0	15	0.15	N	22
268	**Whitecurrants**, *raw*	0	Tr	0	(0.10)	(0.04)	(0.06)	(0.1)	0.2	(0.05)	0	N	(0.06)	(2.6)	(40)
269	*stewed with sugar*	0	Tr	0	(0.08)	(0.02)	(0.04)	(0.1)	0.2	(0.03)	0	N	(0.04)	(1.5)	(23)
270	*stewed without sugar*	0	Tr	0	(0.09)	(0.03)	(0.04)	(0.1)	0.2	(0.03)	0	N	(0.04)	(1.7)	(26)

Fruit juices

14-271 to 14-280

Composition of food per 100g

No. 14-	Food	Description and main data sources	Edible proportion	Water g	Total nitrogen g	Protein g	Fat g	Carbo-hydrate g	Energy value kcal	Energy value kJ
271	**Apple juice**, unsweetened	10 samples; bottles and cartons	1.00	88.0	0.01	0.1	0.1	9.9	38	164
272	**Apple juice concentrate**, unsweetened	10 samples, 68.6 Brix; imported commercial concentrate	1.00	31.4	0.08	0.5	0.6	57.6	223	952
273	**Grape juice**, unsweetened	10 samples, 6 brands; red and white juice	1.00	85.4	0.05	0.3	0.1	11.7	46	196
274	**Grape juice concentrate**	5 samples, red and white juice, 65.6 Brix; imported commercial concentrate	1.00	34.4	0.16	1.0	0.4	58.7	228	971
275	**Grapefruit juice**, unsweetened	50 samples; cartons, canned, bottled and frozen[a]	1.00	89.4	0.07	0.4	0.1	8.3	33	140
276	**Grapefruit juice concentrate**, unsweetened	8 samples, 50.4 Brix; imported commercial concentrate	1.00	49.6	0.35	2.2	0.5	40.8	166	709
277	**Lemon juice**, fresh	Analysis and literature sources	1.00	91.4	0.05	0.3	Tr	1.6	7	31
278	fresh, *weighed as whole fruit*	Calculated from 277	0.35	32.0	0.02	0.1	Tr	0.6	2	11
279	**Lime juice**, fresh	Ref. 3	1.00	90.2	0.06	0.4	0.1	(1.6)	9	36
280	**Mango juice**, canned	Refs. 11, 10	1.00	87.8	0.02	0.1	0.2	(9.8)	39	166

[a] Frozen samples were diluted as per manufacturers' instructions prior to analysis

Fruit juices

Carbohydrate fractions, g per 100g

No. 14-	Food	Starch	Total sugars	Gluc	Fruct	Sucr	Malt	Lact	Dietary fibre Southgate method	Dietary fibre Englyst method	Cellulose	Non-cellulosic polysaccharide Soluble	Non-cellulosic polysaccharide Insoluble	Lignin
271	**Apple juice**, unsweetened	0	9.9	2.6	6.3	1.1	0	0	Tr	Tr	Tr	Tr	Tr	0
272	**Apple juice concentrate**, unsweetened	0	57.6	13.2	33.9	10.4	0	0	Tr	Tr	Tr	Tr	Tr	0
273	**Grape juice**, unsweetened	0	11.7	5.5	6.2	Tr	0	0	0	0	0	0	0	0
274	**Grape juice concentrate**	0	58.7	28.3	29.8	0.6	0	0	0	0	0	0	0	0
275	**Grapefruit juice**, unsweetened	0	8.3	3.0	3.3	2.0	0	0	Tr	Tr	Tr	Tr	Tr	0
276	**Grapefruit juice concentrate**, unsweetened	0	40.8	13.1	12.8	14.9	0	0	Tr	Tr	Tr	Tr	Tr	0
277	**Lemon juice**, fresh	0	1.6	0.5	0.9	0.2	0	0	0.1	0.1	Tr	Tr	0.1	0
278	fresh, *weighed as whole fruit*	0	0.6	0.2	0.3	0.1	0	0	Tr	Tr	Tr	Tr	Tr	0
279	**Lime juice**, fresh	0	(1.6)	(0.6)	(0.6)	(0.4)	0	0	0.1	0.1	Tr	Tr	0.1	0
280	**Mango juice**, canned	(0.2)	(9.6)	(0.5)	(2.1)	(7.0)	0	0	Tr	Tr	Tr	Tr	Tr	0

Fruit juices

Inorganic constituents per 100g

No. 14-	Food	Na	K	Ca	Mg	P	Fe	Cu	Zn	S	Cl	Mn	Se	I
						mg							µg	
271	**Apple juice**, unsweetened	2	110	7	5	6	0.1	Tr	Tr	5	3	Tr	Tr	Tr
272	**Apple juice concentrate**, unsweetened	11	640	35	29	38	1.6	Tr	0.1	29	7	0.5	Tr	Tr
273	**Grape juice**, unsweetened	7	55	19	7	14	0.9	Tr	0.1	6	6	0.1	(1)	N
274	**Grape juice concentrate**	11	160	47	29	40	1.8	0.28	0.2	21	15	0.4	(3)	N
275	**Grapefruit juice**, unsweetened	7	100	14	8	11	0.2	0.01	Tr	4	4	0.2	(1)	N
276	**Grapefruit juice concentrate,** unsweetened	8	560	48	37	60	0.6	0.08	0.1	19	34	0.5	(3)	N
277	**Lemon juice**, fresh	1	130	7	7	8	0.1	0.03	Tr	2	3	Tr	(1)	N
278	fresh, *weighed as whole fruit*	Tr	45	2	2	3	Tr	0.01	Tr	1	1	Tr	Tr	Tr
279	**Lime juice**, fresh	1	110	9	6	7	Tr	0.03	0.1	(2)	(3)	Tr	N	N
280	**Mango juice**, canned	9	18	2	N	22	1.5	N	N	N	N	N	N	N

Fruit juices

No. 14-	Food	Retinol µg	Carotene µg	Vitamin D µg	Vitamin E mg	Thiamin mg	Ribo-flavin mg	Niacin mg	Trypt 60 mg	Vitamin B6 mg	Vitamin B12 µg	Folate µg	Panto-thenate mg	Biotin µg	Vitamin C mg
271	**Apple juice**, unsweetened	0	Tr	0	Tr	0.01	0.01	0.1	Tr	0.02	0	4	0.04	0.8	14
272	**Apple juice concentrate**, unsweetened	0	Tr	0	0.02	0.01	0.02	0.7	0.1	0.07	0	1	0.22	2.0	49
273	**Grape juice**, unsweetened	0	Tr	0	Tr	Tr	0.01	0.1	Tr	0.04	0	1	0.03	0.7	Tr
274	**Grape juice concentrate**	0	Tr	0	Tr	0.03	0.03	0.6	0.1	0.23	0	9	0.14	3.0	1
275	**Grapefruit juice**, unsweetened	0	1	0	0.19	0.04	0.01	0.2	Tr	0.02	0	6	0.08	1	31[a]
276	**Grapefruit juice concentrate**, unsweetened	0	1	0	0.70	0.28	0.06	1.0	0.2	0.18	0	38	0.46	2.0	190
277	**Lemon juice**, fresh	0	12	0	N	0.03	0.01	0.1	Tr	0.05	0	13	0.10	0.3	36[b]
278	fresh, *weighed as whole fruit*	0	4	0	N	0.01	Tr	Tr	Tr	0.02	0	5	0.03	0.1	13
279	**Lime juice**, fresh	0	6	0	N	0.02	0.01	0.1	Tr	0.04	0	N	0.14	N	38
280	**Mango juice**, canned	0	210	0	(1.05)	0.01	0.01	0.5	Tr	N	0	N	N	N	(25)

[a] On analysis of samples, the vitamin C content varied according to packaging and length of storage

[b] Commercial lemon juice contains 11mg vitamin C per 100g

Fruit juices *continued*

Composition of food per 100g

No. Food	Description and main data sources	Edible proportion	Water g	Total nitrogen g	Protein g	Fat g	Carbo-hydrate g	Energy value kcal	Energy value kJ
14-									
281 **Orange juice**, freshly squeezed	Strained juice from fresh oranges	1.00	87.7	0.10	0.6	Tr	8.1	33	140
282 freshly squeezed, *weighed as whole fruit*	Calculated from 281	0.46	40.3	0.05	0.3	Tr	3.7	15	64
283 unsweetened	60 samples; cartons, canned, bottled and frozen[a]	1.00	89.2	0.08	0.5	0.1	8.8	36	153
284 **Orange juice concentrate,** unsweetened	17 samples, 58.4 Brix; imported commercial concentrate	1.00	41.6	0.46	2.9	0.5	44.9	185	786
285 **Passion fruit juice**	Refs. 3, 4	1.00	87.3	0.13	0.8	0.1	10.7	47	189
286 **Pineapple juice**, unsweetened	18 samples, cartons only	1.00	87.8	0.05	0.3	0.1	10.5	41	177
287 **Pineapple juice concentrate,** unsweetened	13 samples, 56.5 Brix; imported commercial concentrate	1.00	43.5	0.21	1.3	0.1	47.5	184	786
288 **Pomegranate juice**, fresh	Juice from fresh fruit	0.56	85.4	0.03	0.2	Tr	11.6	44	189
289 **Prune juice**	4 bottled samples; 2 brands	1.00	79.7	0.08	0.5	0.1	14.4	57	243

[a] Frozen samples were diluted as per manufacturers' instructions prior to analysis

Fruit juices continued

Carbohydrate fractions, g per 100g

No. 14-	Food	Starch	Total sugars	Gluc	Fruct	Sucr	Malt	Lact	Dietary fibre Southgate method	Dietary fibre Englyst method	Cellulose	Non-cellulosic polysaccharide Soluble	Non-cellulosic polysaccharide Insoluble	Lignin
281	**Orange juice**, freshly squeezed	0	8.1	2.0	2.2	4.0	0	0	0.1	0.1	Tr	0.1	Tr	0
282	freshly squeezed, *weighed as whole fruit*	0	3.7	0.9	1.0	1.8	0	0	Tr	Tr	Tr	Tr	Tr	0
283	unsweetened	0	8.8	2.8	2.9	3.1	0	0	0.1	0.1	Tr	0.1	Tr	0
284	**Orange juice concentrate**, unsweetened	0	44.9	11.7	12.3	20.9	0	0	Tr	Tr	Tr	Tr	Tr	0
285	**Passion fruit juice**	0	*10.7*	*4.1*	*3.5*	*3.1*	0	0	Tr	Tr	Tr	Tr	Tr	0
286	**Pineapple juice**, unsweetened	0	10.5	2.9	2.9	4.7	0	0	Tr	Tr	Tr	Tr	Tr	0
287	**Pineapple juice concentrate**, unsweetened	0	47.5	13.0	12.4	22.2	0	0	Tr	Tr	Tr	Tr	Tr	0
288	**Pomegranate juice**, fresh	0	11.6	6.3	5.1	0.2	0	0	Tr	Tr	Tr	Tr	Tr	0
289	**Prune juice**	0	14.4	10.0	4.4	Tr	0	0	Tr	Tr	Tr	Tr	Tr	0

Fruit juices *continued*

Inorganic constituents per 100g

No. 14-	Food	Na	K	Ca	Mg	P	Fe	Cu	Zn	S	Cl	Mn	Se	I
							mg						µg	
281	**Orange juice**, freshly squeezed	2	180	12	12	22	0.3	Tr	Tr	5	1	0.1	(1)	(2)
282	freshly squeezed, *weighed as whole fruit*	1	83	5	5	10	0.1	Tr	Tr	2	Tr	Tr	Tr	(1)
283	unsweetened	10	150	10	8	13	0.2	Tr	Tr	5	9	0.1	(1)	(2)
284	**Orange juice concentrate**, unsweetened	10	880	36	46	83	0.4	0.11	0.2	(27)	17	0.1	(5)	(11)
285	**Passion fruit juice**	(19)	(200)	7	(29)	21	0.5	N	(0.8)	N	N	N	N	N
286	**Pineapple juice**, unsweetened	8	53	8	6	1	0.2	0.02	0.1	N	15	0.7	Tr	Tr
287	**Pineapple juice concentrate**, unsweetened	14	600	44	46	29	1.1	0.12	0.3	N	68	4.2	Tr	Tr
288	**Pomegranate juice**, fresh	1	200	3	3	8	0.2	0.07	N	4	53	N	N	N
289	**Prune juice**	12	210	14	11	19	0.7	0.04	0.2	N	5	0.1	Tr	3

Fruit juices *continued*

No. 14-	Food	Retinol µg	Carotene µg	Vitamin D µg	Vitamin E mg	Thiamin mg	Ribo-flavin mg	Niacin mg	Trypt 60 mg	Vitamin B6 mg	Vitamin B12 µg	Folate µg	Panto-thenate mg	Biotin µg	Vitamin C mg
281	**Orange juice**, fresh y squeezed	0	(17)	0	(0.17)	0.08	0.02	0.2	0.1	(0.07)	0	28	(0.13)	(1.0)	48
282	freshly squeezed, *weighed as whole fruit*	0	(8)	0	(0.08)	0.04	0.01	0.1	Tr	(0.03)	0	13	(0.06)	(0.5)	22
283	unsweetened	0	17	0	0.17	0.08	0.02	0.2	0.1	0.07	0	20	0.13	1.0	39[a]
284	**Orange juice concentrate,** unsweetened	0	170	0	0.68	0.31	0.13	1.3	0.3	0.25	0	90	0.73	5.4	210
285	**Passion fruit juice**	0	800	0	N	0.01	0.07	0.7	0.1	N	0	N	N	N	21
286	**Pineapple juice**, unsweetened	0	8	0	0.03	0.06	0.01	0.1	0.1	0.05	0	8	0.07	Tr	11[a]
287	**Pineapple juice concentrate,** unsweetened	0	16	0	0.14	0.23	0.05	0.6	0.3	0.25	0	40	0.31	1.0	82
288	**Pomegranate juice,** fresh	0	(33)	0	N	0.02	0.03	0.2	Tr	(0.31)	0	N	(0.57)	N	8
289	**Prune juice**	0	N	0	N	0.01	0.05	0.6	0.1	0.14	0	Tr	N	N	Tr

[a] On analysis of samples, the vitamin C content varied according to packaging and length of storage

Nuts

and

Seeds

Nuts and seeds

No. 14-	Food	Description and main data sources	Edible proportion	Water g	Total nitrogen g	Protein g	Fat g	Carbo-hydrate g	Energy value kcal	kJ
801	**Almonds**	10 blanched samples, flaked and ground	1.00	4.2	4.07	21.1	55.8	6.9	612	2534
802	*weighed with shells*	Calculated from *801*	0.37	1.5	1.51	7.8	20.6	2.5	229	935
803	*toasted*	Ref. 5 and calculation from *801*	1.00	2.6	4.14	21.2	56.7	7.0	621	2570
804	**Barcelona nuts**	Kernel only	1.00	5.7	2.06	10.9	64.0	5.2	639	2637
805	*weighed with shells*	Calculated from *804*	0.62	3.5	1.28	6.8	39.7	3.2	397	1636
806	**Betel nuts**	Refs. 11, 6	1.00	11.5	0.98	5.2	10.2	56.7	339	1430
807	**Bombay mix**	20 samples; savoury mix of gram flour, assorted peas, lentils, nuts and seeds	1.00	3.5	3.01	18.8	32.9	35.1	503	2099
808	**Brazil nuts**	10 samples, kernel only	1.00	2.8	2.61	14.1	68.2	3.1	682	2813
809	*weighed with shells*	Calculated from *808*	0.46	1.3	1.20	6.5	31.4	1.4	314	1295
810	**Breadnut seeds**	Ref. 5	1.00	6.5	1.63	8.6	1.7	73.8	367	1536
811	**Cashew nuts**, *plain*	20 samples, whole and broken kernels	1.00	4.4	3.34	17.7	48.2	18.1	573	2374
812	*roasted and salted*	10 samples, kernels only	1.00	2.4	3.87	20.5	50.9	18.8	611	2533
813	**Chestnuts**	Analysis and literature sources; kernel only	1.00	51.7	0.37	2.0	2.7	36.6	170	719
814	*weighed with shells*	Calculated from *813*	0.83	42.9	0.31	1.6	2.2	30.4	140	595
815	*dried*	Ref. 5 and calculation from *813*	1.00	9.0	0.69	3.7	5.1	69.0	319	1356

No. 14-	Food	Starch	Total sugars	Gluc	Fruct	Sucr	Malt	Lact	Dietary fibre Southgate method	Dietary fibre Englyst method	Cellulose	Non-cellulosic polysaccharide Soluble	Non-cellulosic polysaccharide Insoluble	Lignin
801	**Almonds**	2.7	4.2	Tr	Tr	4.2	0	0	(12.9)	(7.4)	(1.9)	(1.1)	(4.4)	N
802	*weighed with shells*	1.0	1.5	Tr	Tr	1.5	0	0	(4.8)	(2.7)	(0.7)	(0.4)	(1.6)	N
803	*toasted*	2.7	4.3	Tr	Tr	4.3	0	0	(13.3)	(7.5)	(1.9)	(1.1)	(4.5)	N
804	**Barcelona nuts**	1.8	3.4	0	0	(3.4)	0	0	9.3	N	N	N	N	N
805	*weighed with shells*	1.1	2.1	0	0	(2.1)	0	0	5.8	N	N	N	N	N
806	**Betel nuts**	N	N	0	0	N	0	0	N	N	N	N	N	N
807	**Bombay mix**	32.8	2.3	0.1	0.1	2.2	0	0	N	6.2	0.7	N	N	N
808	**Brazil nuts**	0.7	2.4	0	0	2.4	0	0	8.1	4.3	1.6	1.3	1.4	N
809	*weighed with shells*	0.3	1.1	0	0	1.1	0	0	3.7	2.0	0.7	0.6	0.6	N
810	**Breadnut seeds**	N	N	N	N	N	0	0	N	N	N	N	N	N
811	**Cashew nuts**, *plain*	13.5	4.6	0	0	4.6	0	0	N	3.2	0.6	1.6	1.0	N
812	*roasted and salted*	13.2	5.6	0	0	5.6	0	0	N	3.2	0.4	1.7	1.1	N
813	**Chestnuts**	29.6	7.0	Tr	Tr	7.0	0	0	6.1	4.1	1.1	1.3	1.7	N
814	*weighed with shells*	24.6	5.8	Tr	Tr	5.8	0	0	5.1	3.4	0.9	1.1	1.4	N
815	*dried*	55.8	13.2	Tr	Tr	13.2	0	0	11.5	7.7	2.1	2.5	3.2	N

Nuts and seeds

Inorganic constituents per 100g

No. 14-	Food	Na	K	Ca	Mg	P	Fe	Cu	Zn	S	Cl	Mn	Se	I
							mg						µg	
801	**Almonds**	14	780	240	270	550	3.0	1.00	3.2	150	18	1.7	4	2
802	*weighed with shells*	5	290	89	100	200	1.1	0.37	1.2	55	7	0.6	1	1
803	*toasted*	14	790	240	270	560	3.1	1.02	3.3	150	18	1.7	4	2
804	**Barcelona nuts**	3	940	170	200	300	3.0	0.96	N	180	34	N	N	N
805	*weighed with shells*	2	580	110	120	190	1.9	0.59	N	110	21	N	N	N
806	**Betel nuts**	76	450	400	N	89	4.9	N	N	N	N	N	N	N
807	**Bombay mix**	770	770	58	100	290	3.8	0.62	2.5	N	1410	1.4	N	N
808	**Brazil nuts**	3	660	170	410	590	2.5	1.76	4.2	290	57	1.2	1530[a]	20
809	*weighed with shells*	1	300	78	190	270	1.1	0.81	1.9	130	26	0.5	700	9
810	**Breadnut seeds**	N	N	94	N	180	4.6	N	N	N	N	N	N	N
811	**Cashew nuts**, *plain*	15	710	32	270	560	6.2	2.11	5.9	N	17	1.7	29	(11)
812	*roasted and salted*	290	730	35	250	510	6.2	2.04	5.7	N	490	1.8	34	11
813	**Chestnuts**	11	500	46	33	74	0.9	0.23	0.5	29	15	0.5	Tr	N
814	*weighed with shells*	9	410	38	27	61	0.7	0.19	0.4	24	12	0.4	Tr	N
815	*dried*	21	940	87	62	140	1.7	0.43	0.9	55	28	0.9	Tr	N

[a] Selenium can range from 230 to 5300µg per 100g

90

No. 14-	Food	Retinol µg	Carotene µg	Vitamin D µg	Vitamin E mg	Thiamin mg	Ribo-flavin mg	Niacin mg	Trypt 60 mg	Vitamin B6 mg	Vitamin B12 µg	Folate µg	Panto-thenate mg	Biotin µg	Vitamin C mg
801	**Almonds**	0	0	0	23.96	0.21	0.75	3.1	3.4	0.15	0	48	0.44	64.0	0
802	*weighed with shells*	0	0	0	8.86	0.08	0.28	1.1	1.3	0.05	0	18	0.16	24.0	0
803	*toasted*	0	0	0	24.36	0.13	0.57	2.6	3.5	0.09	0	36	0.25	49.0	0
804	**Barcelona nuts**	0	0	0	N	0.11	N	N	3.1	N	0	N	N	N	Tr
805	*weighed with shells*	0	0	0	N	0.07	N	N	1.9	N	0	N	N	N	Tr
806	**Betel nuts**	0	0	0	N	0.19	0.52	1.1	N	N	0	N	N	N	Tr
807	**Bombay mix**	0	Tr	0	4.71	0.38	0.10	4.3	3.5	0.54	0	N	1.19	24	Tr
808	**Brazil nuts**	0	0	0	7.18	0.67	0.03	0.3	3.0	0.31	0	21	0.41	11.0	0
809	*weighed with shells*	0	0	0	3.30	0.31	0.01	0.1	1.4	0.14	0	10	0.19	5.0	0
810	**Breadnut seeds**	0	(130)	0	N	0.03	0.14	2.1	2.7	N	0	N	N	N	47
811	**Cashew nuts**, *plain*	0	6	0	0.85	0.69	0.14	1.2	4.5	0.49	0	67	1.06	12.7	0
812	*roasted and salted*	0	6	0	1.30	0.41	0.16	1.3	5.2	0.43	0	68	1.08	13.0	0
813	**Chestnuts**	0	0	0	1.20	0.14	0.02	0.5	0.4	0.34	0	N	0.49	1.4	Tr
814	*weighed with shells*	0	0	0	0.99	0.12	0.02	0.4	0.3	0.28	0	N	0.41	1.2	Tr
815	*dried*	0	0	0	2.26	(0.26)	(0.04)	(0.9)	0.8	(0.64)	0	N	(0.92)	(2.6)	Tr

Nuts and seeds *continued*

14-816 to 14-830

Composition of food per 100g

No. 14-	Food	Description and main data sources	Edible proportion	Water g	Total nitrogen g	Protein g	Fat g	Carbo-hydrate g	Energy value kcal	kJ
816	**Coconut**, *fresh*	Flesh from kernel, no shell	0.70[a]	45.0	0.61	3.2	36.0	3.7	351	1446
817	*creamed block*	7 samples, 2 brands; block of dried kernel	1.00	2.5	1.14	6.0	68.8	7.0	669	2760
818	*desiccated*	Analytical and literature sources	1.00	2.3	1.05	5.6	62.0	6.4	604	2492
819	**Coconut cream**	Literature sources. Liquid from grated coconut kernel	1.00	53.9	0.75	4.0	34.7	5.9	350	1446
820	**Coconut milk**	Analysis and literature sources; drained fluid from fresh coconut	1.00	92.2	0.06	0.3	0.3	4.9	22	95
821	**Hazelnuts**	10 samples, kernel only	1.00	4.6	2.66	14.1	63.5	6.0	650	2685
822	*weighed with shells*	Calculated from 821	0.38	1.7	1.01	5.4	24.1	2.3	247	1020
823	**Macadamia nuts**, salted	8 samples	1.00	1.3	1.49	7.9	77.6	4.8	748	3082
824	**Marzipan**, *homemade*	Recipe[b]	1.00	10.2	1.98	10.4	25.8	50.2	461	1933
825	*retail*	10 samples, white and yellow	1.00	7.9	1.02	5.3	14.4	67.6	404	1705
826	**Melon seeds**	Literature sources	1.00	6.1	5.38	28.5	47.7	9.9	583	2418
827	**Mixed nuts**	Calculated as peanuts 67%, almonds 17%, cashews 8% and hazelnuts 7%	1.00	2.5	4.27	22.9	54.1	7.9	607	2515
828	**Mixed nuts and raisins**	Calculated as raisins 37%, peanuts 36%, hazelnuts 12%, brazil nuts 11% and almonds 4%	1.00	8.2	2.60	14.1	34.1	31.5	481	2004
829	**Peanut butter**, smooth	10 samples, 3 brands	1.00	1.1	4.17	22.6	53.7	13.1	623	2581
830	*wholegrain*	7 samples, 2 brands; peanuts, oil and salt only	1.00	0.7	4.60	24.9	53.1	7.7	606	2511

[a] Weighed whole, including shell = 0.50

[b] Calculated from recipe containing 300g ground almonds, 300g sugar, 50g egg and 20ml lemon juice

Nuts and seeds *continued*

Carbohydrate fractions, g per 100g

No. 14-	Food	Starch	Total sugars	Gluc	Fruct	Sucr	Malt	Lact	Dietary fibre Southgate method	Dietary fibre Englyst method	Cellulose	Non-cellulosic polysaccharide Soluble	Non-cellulosic polysaccharide Insoluble	Lignin
816	Coconut, fresh	0	3.7	Tr	Tr	3.7	0	0	12.2	7.3	0.8	1.0	5.5	N
817	creamed block	0	7.0	Tr	0.1	6.9	0	0	N	N	N	N	N	N
818	desiccated	0	6.4	Tr	0.8	5.6	0	0	21.1	13.7	1.5	1.4	10.8	N
819	Coconut cream	0	5.9	Tr	0.1	5.8	0	0	N	N	N	N	N	Tr
820	Coconut milk	0	4.9	0.3	Tr	4.6	0	0	Tr	Tr	Tr	Tr	Tr	0
821	Hazelnuts	2.0	4.0	0.2	0.1	3.7	0	0	8.9	6.5	2.2	2.5	1.8	N
822	weighed with shells	0.8	1.5	0.1	Tr	1.4	0	0	3.4	2.5	0.8	0.9	0.7	N
823	Macadamia nuts, salted	0.8	4.0	0.1	0.1	3.8	0	0	N	5.3	1.4	1.9	2.0	N
824	Marzipan, homemade	1.2	48.9	Tr	Tr	48.9	0	0	(5.8)	(3.3)	(0.9)	(0.5)	(2.0)	N
825	retail	Tr	67.6	2.7	1.1	62.2	1.6	0	(3.2)	(1.9)	(0.5)	(0.3)	(1.1)	N
826	Melon seeds	N	N	N	N	N	0	0	N	N	N	N	N	N
827	Mixed nuts	3.9	4.0	Tr	Tr	4.0	0	0	7.5	6.0	1.7	1.8	2.7	N
828	Mixed nuts and raisins	2.7	28.8	12.8	12.9	3.1	0	0	7.4	4.5	1.5	1.5	1.4	N
829	Peanut butter, smooth	6.4	6.7	0	0	6.7	0	0	6.8	5.4	1.6	1.6	2.2	N
830	wholegrain	3.3	4.4	0	0	4.4	0	0	(6.9)	(6.0)	(1.7)	(1.9)	(2.5)	N

Nuts and seeds *continued*

Inorganic constituents per 100g

No. 14-	Food	Na	K	Ca	Mg	P	Fe (mg)	Cu	Zn	S	Cl	Mn	Se (μg)	I
816	**Coconut**, *fresh*	17	370	13	41	94	2.1	0.32	0.5	44	110	1.0	(1)	(1)
817	*creamed block*	30	650	23	73	170	3.7	0.56	0.9	78	190	1.8	(2)	(2)
818	*desiccated*	28	660	23	90	160	3.6	0.55	0.9	76	200	1.8	(3)	(3)
819	**Coconut cream**	4	330	11	46	120	2.3	0.40	1.0	37	92	1.3	(1)	(1)
820	**Coconut milk**	110	280	29	30	30	0.1	0.04	0.1	24	180	N	N	N
821	**Hazelnuts**	6	730	140	160	300	3.2	1.23	2.1	120	18	4.9	Tr	17
822	*weighed with shells*	2	280	53	61	110	1.2	0.47	0.8	46	7	1.9	Tr	6
823	**Macadamia nuts**, *salted*	280	300	47	100	200	1.6	0.43	1.1	N	390	5.5	7	N
824	**Marzipan**, *homemade*	16	360	110	120	260	1.5	0.46	1.6	81	20	0.8	3	5
825	*retail*	20	160	66	68	130	0.9	0.24	0.8	38	23	0.4	1	Tr
826	**Melon seeds**	99	650	71	510	690	7.6	(2.39)	(4.0)	N	N	(2.3)	N	N
827	**Mixed nuts**	300	790	78	200	430	2.1	0.79	3.1	280	490	2.1	5	12
828	**Mixed nuts and raisins**	24	820	84	160	310	2.3	0.82	2.4	200	14	1.6	170	N
829	**Peanut butter**, *smooth*	350	700	37	180	330	2.1	0.70	3.0	330	500	1.7	3	N
830	*wholegrain*	370	680	47	18	370	2.5	0.68	3.5	360	540	1.8	4	N

No. 14-	Food	Retinol µg	Carotene µg	Vitamin D µg	Vitamin E mg	Thiamin mg	Ribo-flavin mg	Niacin mg	Trypt 60 mg	Vitamin B6 mg	Vitamin B12 µg	Folate µg	Panto-thenate mg	Biotin µg	Vitamin C mg
816	**Coconut**, *fresh*	0	0	0	0.73	0.04	0.01	0.5	0.6	0.05	0	26	0.30	N	3
817	*creamed block*	0	0	0	1.40	(0.03)	(0.05)	(0.9)	1.2	N	0	(9)	(0.50)	N	0
818	*desiccated*	0	0	0	1.26	0.03	0.05	0.9	1.1	(0.09)	0	9	0.50	N	0
819	**Coconut cream**	0	0	0	0.70	0.03	Tr	0.9	0.8	N	0	N	N	N	3
820	**Coconut milk**	0	0	0	Tr	0.03	0.06	0.1	0.1	0.03	0	N	0.04	N	2
821	**Hazelnuts**	0	0	0	24.98	0.43	0.16	1.1	4.0	0.59	0	72	1.51	76.0	0
822	*weighed with shells*	0	0	0	9.49	0.16	0.06	0.4	1.5	0.22	0	27	0.57	29.0	0
823	**Macadamia nuts**, *salted*	0	0	0	1.49	0.28	0.06	1.6	1.7	0.28	0	N	0.61	6.0	0
824	**Marzipan**, *homemade*	14	Tr	0.1	10.82	0.10	0.37	1.4	1.8	0.08	0.2	25	0.33	30.2	1
825	*retail*	0	0	0	6.18	(0.05)	(0.19)	(0.7)	0.9	(0.04)	0	(12)	(0.11)	(16.0)	0
826	**Melon seeds**	0	Tr	0	N	0.17	0.15	2.1	9.7	N	0	58	N	N	Tr
827	**Mixed nuts**	0	Tr	0	6.44	0.22	0.22	9.9	4.9	0.53	0	54	1.42	86.4	0
828	**Mixed nuts and raisins**	0	4	0	N	0.59	0.11	5.5	3.0	0.42	0	56	1.26	39.5	Tr
829	**Peanut butter**, *smooth*	0	0	0	4.99	0.17	0.09	12.5	4.9	0.58	0	53	1.56	94.0	0
830	*wholegrain*	0	0	0	N	(0.18)	(0.10)	(13.6)	5.4	(0.63)	0	(52)	(1.70)	(102.0)	0

Nuts and seeds *continued*

Composition of food per 100g

No. 14-	Food	Description and main data sources	Edible proportion	Water g	Total nitrogen g	Protein g	Fat g	Carbo-hydrate g	Energy value kcal	kJ
831	**Peanuts,** *plain*	10 samples, kernel only	1.00	6.3	4.73	25.6	46.1	12.5	564	2341
832	*plain, weighed with shells*	Calculated from 831	0.69	4.3	3.26	17.7	31.8	8.6	389	1615
833	*dry roasted*	10 samples, 5 brands	1.00	1.8	4.71	25.5	49.8	10.3	589	2441
834	*roasted and salted*	20 samples	1.00	1.9	4.53	24.5	53.0	7.1	602	2491
835	**Peanuts, raisins and chocolate chips**	Calculated as raisins 39%, peanuts 37% and chocolate chips 24%	1.00	8.0	2.21	12.3	24.5	45.9	441	1849
836	**Peanuts and raisins**	Calculated as peanuts 56% and raisins 44%	1.00	9.3	2.80	15.3	26.0	37.5	435	1820
837	**Pecan nuts**	9 samples, kernel only	1.00	3.7	1.74	9.2	70.1	5.8	689	2843
838	*weighed with shells*	Calculated from 837	0.49	1.8	0.85	4.5	34.3	2.8	337	1390
839	**Pine nuts**	20 samples, pine kernels	1.00	2.7	2.64	14.0	68.6	4.0	688	2840
840	**Pistachio nuts,** *roasted and salted*	10 samples, kernel only	1.00	2.1	3.38	17.9	55.4	8.2	601	2485
841	*roasted and salted, weighed with shells*	Calculated from 840	0.55	1.1	1.86	9.9	30.5	4.6	331	1370
842	**Pumpkin seeds**	Analysis and literature sources	1.00	5.6	4.61	24.4	45.6	15.2	569	2360
843	**Quinoa**	Literature sources	1.00	11.5	2.21	13.8	5.0	55.7[a]	309	1311
844	**Sesame seeds**	10 samples, with and without hulls	1.00	4.6	3.44	18.2	58.0	0.9	598	2470
845	**Sunflower seeds**	Analysis and literature sources	1.00	4.4	3.74	19.8	47.5	18.6[a]	581	2410
846	*toasted*	Calculated from 845	1.00	1.0	3.86	20.5	49.2	19.3[a]	602	2497

[a] Including oligosaccharides

Nuts and seeds *continued*

Carbohydrate fractions, g per 100g

No. Food 14-	Starch	Total sugars	Individual sugars					Dietary fibre		Fibre fractions			
			Gluc	Fruct	Sucr	Malt	Lact	Southgate method	Englyst method	Cellulose	Non-cellulosic polysaccharide Soluble	Insoluble	Lignin
831 **Peanuts, plain**	6.3	6.2	0	0	6.2	0	0	7.3	6.2	2.0	1.9	2.3	N
832 *plain, weighed with shells*	4.3	4.3	0	0	4.3	0	0	5.0	4.3	1.4	1.3	1.6	N
833 *dry roasted*	6.5	3.8	0	0	3.8	0	0	7.4	6.4	1.8	2.0	2.7	N
834 *roasted and salted*	3.3	3.8	0	0	3.8	0	0	6.9	6.0	1.7	1.9	2.5	N
835 **Peanuts, raisins and chocolate chips**	3.0	42.9	13.5	13.6	14.8	0	1.1	5.1	3.1	1.1	1.1	0.9	N
836 **Peanuts and raisins**	3.5	34.0	15.2	15.3	3.5	0	0	6.8	4.4	1.5	1.5	1.4	N
837 **Pecan nuts**	1.5	4.3	0.3	0.3	3.7	0	0	N	4.7	1.2	1.5	2.0	N
838 *weighed with shells*	0.7	2.1	0.1	0.1	1.8	0	0	N	2.3	0.6	0.7	1.0	N
839 **Pine nuts**	0.1	3.9	0.1	0.1	3.7	0	0	N	1.9	N	N	N	N
840 **Pistachio nuts,** *roasted and salted*	2.5	5.7	Tr	Tr	5.7	0	0	N	6.1	1.3	2.7	2.1	N
841 *roasted and salted, weighed with shells*	1.4	3.2	Tr	Tr	3.2	0	0	N	3.3	0.7	1.5	1.1	N
842 **Pumpkin seeds**	14.1	1.1	0	0	1.1	0	0	N	5.3	1.1	1.7	2.5	N
843 **Quinoa**	47.6	6.1a	2.7	0.6	2.8	0	0	N	N	4.4	N	N	N
844 **Sesame seeds**	0.5	0.4	0.1	0.1	0.2	0	0	N	7.9	N	N	N	N
845 **Sunflower seeds**	16.3	1.7a	0	0	1.7	0	0	N	6.0	1.4	1.8	2.8	N
846 *toasted*	16.7	1.8a	0	0	1.8	0	0	N	6.2	1.4	1.9	2.9	N

a Not including oligosaccharides

Nuts and seeds continued

Inorganic constituents per 100g

No. 14-	Food	Na	K	Ca	Mg	P	Fe	Cu	Zn	S	Cl	Mn	Se	I
							mg						µg	
831	**Peanuts,** *plain*	2	670	60	210	430	2.5	1.02	3.5	380	7	2.1	3	20
832	*plain, weighed with shells*	1	460	41	140	300	1.7	0.70	2.4	260	5	1.4	2	14
833	*dry roasted*	790	730	52	190	420	2.1	0.64	3.3	380	1140	2.2	3	19
834	*roasted and salted*	400	810	37	180	410	1.3	0.54	2.9	360	660	1.9	4	19
835	**Peanuts, raisins and chocolate chips**	52	750	92	100	250	1.9	0.60	1.6	N	70	1.0	2	N
836	**Peanuts and raisins**	27	820	53	130	270	2.1	0.74	2.3	220	7	1.3	2	11
837	**Pecan nuts**	1	520	61	130	310	2.2	1.07	5.3	N	15	4.6	12	N
838	*weighed with shells*	Tr	250	30	64	150	1.1	0.52	2.6	N	7	2.3	6	N
839	**Pine nuts**	1	780	11	270	650	5.6	1.32	6.5	N	41	7.9	N	N
840	**Pistachio nuts,** *roasted and salted*	530	1040	110	130	420	3.0	0.83	2.2	N	810	0.9	(6)	N
841	*roasted and salted, weighed with shells*	290	570	61	71	230	1.7	0.46	1.2	N	450	0.5	(3)	N
842	**Pumpkin seeds**	18	820	39	270	850	10.0	1.57	6.6	N	N	N	(6)	N
843	**Quinoa**	61	780	79	210	230	7.8	0.82	3.3	N	N	N	N	N
844	**Sesame seeds**	20	570	670	370	720	10.4	1.46	5.3	N	10	1.5	N	N
845	**Sunflower seeds**	3	710	110	390	640	6.4	2.27	5.1	N	N	2.2	(49)	N
846	*toasted*	3[a]	730	110	400	660	6.6	2.35	5.3	N	N	2.3	(51)	N

[a] Salted toasted sunflower seeds contain approximately 610mg Na per 100g

No. 14-	Food	Retinol µg	Carotene µg	Vitamin D µg	Vitamin E mg	Thiamin mg	Ribo-flavin mg	Niacin mg	Trypt 60 mg	Vitamin B6 mg	Vitamin B12 µg	Folate µg	Panto-thenate mg	Biotin µg	Vitamin C mg
831	**Peanuts**, *plain*	0	0	0	10.09	1.14	0.10	13.8	5.5	0.59	0	110	2.66	72.0	0
832	*plain, weighed with shells*	0	0	0	6.97	0.79	0.07	9.5	3.8	0.41	0	76	1.83	49.7	0
833	*dry roasted*	0	0	0	1.11	0.18	0.13	13.1	5.5	0.54	0	66	1.59	130.0	0
834	*roasted and salted*	0	0	0	0.66	0.18	0.10	13.6	5.3	0.63	0	52	1.70	102.0	0
835	**Peanuts, raisins and chocolate chips**	Tr	14	Tr	N	0.49	0.11	5.4	2.4	0.33	Tr	46	1.30	28.1	Tr
836	**Peanuts and raisins**	0	5	0	5.65	0.69	0.08	8.0	3.1	0.44	0	65	1.56	41.2	Tr
837	**Pecan nuts**	0	50	0	4.34	0.71	0.15	1.4	4.1	0.19	0	39	1.71	N	0
838	*weighed with shells*	0	25	0	2.12	0.35	0.07	0.7	2.0	0.09	0	19	0.84	N	0
839	**Pine nuts**	0	10	0	13.65	0.73	0.19	3.8	3.1	N	0	N	N	N	Tr
840	**Pistachio nuts**, *roasted and salted*	0	130	0	4.16	0.70	0.23	1.7	3.9	N	0	58	N	N	0
841	*roasted and salted, weighed with shells*	0	71	0	2.28	0.39	0.13	0.9	2.2	N	0	32	N	N	0
842	**Pumpkin seeds**	0	(230)	0	N	0.23	0.32	1.7	7.1	N	0	N	N	N	0
843	**Quinoa**	0	N	0	N	0.20	0.40	2.9	1.9	N	0	N	N	N	0
844	**Sesame seeds**	0	6	0	2.53	0.93	0.17	5.0	5.4	0.75	0	97	2.14	11.0	0
845	**Sunflower seeds**	0	15	0	37.77	1.60	0.19	4.1	5.0	N	0	N	N	N	0
846	*toasted*	0	15	0	(39.11)	N	N	N	5.1	N	0	N	N	N	0

Nuts and seeds *continued*

Composition of food per 100g

No. 14-	Food	Description and main data sources	Edible proportion	Water g	Total nitrogen g	Protein g	Fat g	Carbohydrate g	Energy value kcal	kJ
847	**Tahini paste**	Ref. 5 and calculation from *844*	1.00	3.1	3.49	18.5	58.9	0.9	607	2508
848	**Tigernuts**	Literature sources	1.00	8.5	0.69	4.3	23.8	45.7	403	1685
849	**Trail mix**	10 samples; mix of nuts and dried fruit	1.00	8.9	1.45	9.1	28.5	37.2	432	1804
850	**Walnuts**	10 samples, kernel only	1.00	2.8	2.77	14.7	68.5	3.3	688	2837
851	*weighed with shells*	Calculated from *850*	0.43	1.2	1.19	6.3	29.4	1.4	295	1217

Carbohydrate fractions, g per 100g

No. Food 14-	Starch	Total sugars	Individual sugars					Dietary fibre		Fibre fractions			
								Southgate method	Englyst method	Cellulose	Non-cellulosic polysaccharide		Lignin
			Gluc	Fruct	Sucr	Malt	Lact				Soluble	Insoluble	
847 **Tahini paste**	0.5	0.4	0.1	0.1	0.2	0	0	N	8.0	N	N	N	N
848 **Tigernuts**	29.6	16.1	0	0	16.1	0	0	(17.4)	11.7	4.3	1.1	6.3	5.7
849 **Trail mix**	0.1	37.1	17.3	16.1	3.4	0.4	0	N	4.3	1.1	N	N	N
850 **Walnuts**	0.7	2.6	0.2	0.2	2.2	0	0	5.9	3.5	1.1	1.5	0.9	N
851 *weighed with shells*	0.3	1.1	0.1	0.1	0.9	0	0	2.5	1.5	0.5	0.6	0.4	N

Nuts and seeds *continued*

Inorganic constituents per 100g

No. 14-	Food	Na	K	Ca	Mg	P	mg Fe	Cu	Zn	S	Cl	Mn	µg Se	I
847	**Tahini paste**	20	580	680	380	730	10.6	1.48	5.4	N	10	1.5	N	N
848	**Tigernuts**	(1)	(14)	48	N	210	3.2	N	N	N	N	N	N	N
849	**Trail mix**	27	620	69	110	210	3.7	0.55	1.5	N	N	1.6	N	N
850	**Walnuts**	7	450	94	160	380	2.9	1.34	2.7	140	24	3.4	19	9
851	*weighed with shells*	3	190	40	69	160	1.2	0.58	1.2	60	10	1.5	8	4

No. 14-	Food	Retinol µg	Carotene µg	Vitamin D µg	Vitamin E mg	Thiamin mg	Ribo-flavin mg	Niacin mg	Trypt 60 mg	Vitamin B_6 mg	Vitamin B_{12} µg	Folate µg	Panto-thenate mg	Biotin µg	Vitamin C mg
847	**Tahini paste**	0	6	0	2.57	0.94	0.17	5.1	4.1	0.76	0	99	2.17	11.0	0
848	**Tigernuts**	0	0	0	N	0.23	0.10	1.1	0.7	N	0	N	N	N	6
849	**Trail mix**	0	47	0	4.53	0.23	0.09	2.0	1.5	N	0	25	N	N	Tr
850	**Walnuts**	0	0	0	3.85	0.40	0.14	1.2	2.8	0.67	0	66	1.60	19.0	0[a]
851	*weighed with shells*	0	0	0	1.66	0.17	0.06	0.5	1.2	0.29	0	28	0.69	8.0	0

[a] Value for ripe walnuts. Unripe walnuts contain 1300 to 3000mg vitamin C per 100g

Appendices

FAT FRACTIONS (g per 100g) AND CHOLESTEROL (mg per 100g)

Total saturated, monounsaturated and polysaturated fatty acids, and cholesterol are given below for foods where total fat is equal to or greater than 1g per 100g. The values refer to the edible portion of the food except where specified. The fat fractions have been calculated from the total fat after applying conversion factors to allow for the non fatty acid material present. The factors applied were 0.956 for nuts and avocado and 0.8 for olives.

No. 14-	Food	Fat	Satd	Mono unsatd	Poly unsatd	Cholest- erol
				Fatty acid totals		
Fruit						
37	**Avocado**, *average*	19.5	4.1	12.1	2.2	0
38	*average, weighed with skin and stone*	13.8	2.9	8.6	1.6	0
39	Fuerte	19.3	3.5	12.9	1.8	0
40	*weighed with skin and stone*	13.7	2.5	9.2	1.3	0
41	Hass	19.7	4.7	11.3	2.6	0
42	*weighed with skin and stone*	14.0	3.3	8.0	1.8	0
173	**Olives**, *in brine*	11.0	1.7	5.7	1.3	0
174	*in brine, weighed with stones*	8.8	1.4	4.6	1.0	0
Nuts and seeds						
801	**Almonds**	55.8	4.7	34.4	14.2	0
802	*weighed with shells*	20.6	1.7	12.7	5.3	0
803	*toasted*	56.7	4.7	35.0	14.5	0
807	**Bombay mix**	32.9	4.0	16.2	11.3	0
808	**Brazil nuts**	68.2	16.4	25.8	23.0	0
809	*weighed with shells*	31.4	7.5	11.9	10.6	0
810	**Breadnut seeds**	1.7	0.5	0.2	0.9	0
811	**Cashew nuts**, *plain*	48.2	9.5	27.8	8.8	0
812	*roasted and salted*	50.9	10.1	29.4	9.1	0
813	**Chestnuts**	2.7	0.5	1.0	1.1	0
814	*weighed with shells*	2.2	0.4	0.8	0.9	0
815	*dried*	5.1	0.9	1.9	2.1	0
816	**Coconut**, *fresh*	36.0	31.0	2.0	0.8	0
817	*creamed block*	68.8	59.3	3.9	1.6	0
818	*desiccated*	62.0	53.4	3.5	1.5	0
819	**Coconut cream**	34.7	29.9	2.0	0.8	0

No. 14-	Food	Fat	Fatty acid totals			Cholest-erol
			Satd	Mono unsatd	Poly unsatd	
821	**Hazelnuts**	63.5	4.7	50.0	5.9	0
822	weighed with shells	24.1	1.8	18.9	2.3	0
823	**Macadamia nuts**, salted	77.6	11.2	60.8	1.6	0
824	**Marzipan**, homemade	25.8	2.3	15.7	6.5	28
825	retail	14.4	1.2	8.9	3.7	0
826	**Melon seeds**	47.7	12.0	8.3	26.2	0
827	**Mixed nuts**	54.1	8.4	28.2	14.8	0
828	**Mixed nuts and raisins**	34.1	5.5	17.8	8.9	0
829	**Peanut butter,** smooth	53.7	11.7	21.3	18.4	0
830	wholegrain	53.1	(9.5)	(24.2)	(16.5)	0
831	**Peanuts**, plain	46.1	8.2	21.1	14.3	0
832	plain, weighed with shells	31.8	5.7	14.5	9.9	0
833	dry roasted	49.8	8.9	22.8	15.5	0
834	roasted and salted	53.0	9.5	24.2	16.5	0
835	**Peanuts, raisins and chocolate chips**	24.5	7.3	10.1	5.7	7
836	**Peanuts and raisins**	26.0	4.6	11.8	8.0	0
837	**Pecan nuts**	70.1	5.7	42.5	18.7	0
838	weighed with shells	34.3	2.8	20.8	9.2	0
839	**Pine nuts**	68.6	4.6	19.9	41.1	0
840	**Pistachio nuts**, roasted and salted	55.4	7.4	27.6	17.9	0
841	roasted and salted, weighed with shells	30.5	4.1	15.2	9.8	0
842	**Pumpkin seeds**	45.6	7.0	11.2	18.3	0
843	**Quinoa**	5.0	0.5	1.4	2.1	0
844	**Sesame seeds**	58.0	8.3	21.7	25.5	0
845	**Sunflower seeds**	47.5	4.5	9.8	31.0	0
846	toasted	49.2	4.7	10.1	32.1	0
847	**Tahini paste**	58.9	8.4	22.0	25.8	0
848	**Tigernuts**	23.8	4.0	16.4	2.2	0
850	**Walnuts**	68.5	5.6	12.4	47.5	0
851	weighed with shells	29.4	2.4	5.3	20.4	0

CAROTENOID FRACTIONS

β-Carotene is the main or only source of vitamin A activity in most fruit. When analysis has shown significant amounts of other carotenoids, these are listed below. The values for cryptoxanthins were often unspecified, the beta form is likely to predominate in most fruit with smaller amounts of the alpha form present. The β-carotene equivalent is the sum of β-carotene and half of any α-carotene or cryptoxanthins present, and the retinol equivalent is one sixth of the β-carotene equivalent. Nuts contain little or no carotene. Values over 100μg have in all cases been rounded to the nearest 5μg.

			Carotenoid fractions, μg per 100g			
		Carotenoid fractions				
		α-carotene	β-carotene	crypto-xanthins	Carotene equiv	Retinol equiv
Fruit						
19	**Apples, eating**, Golden Delicious, *raw*	0	10	10	15	3
20	Golden Delicious, *raw, weighed with core*	0	9	9	13	2
21	Granny Smith, *raw*	0	Tr	10	5	1
22	Granny Smith, *raw, weighed with core*	0	Tr	9	5	1
23	red dessert, *raw*	0	10	10	15	3
24	red dessert, *raw, weighed with core*	0	9	9	13	2
25	**Apricots**, *raw*	2	405	0	405	67
26	*raw, weighed with stones*	2	375	0	375	63
31	dried	9	640	0	645	105
32	dried, stewed with sugar	3	250	0	255	43
33	-, stewed without sugar	3	260	0	265	44
34	canned in syrup	3	150	0	155	25
36	ready-to-eat	8	540	0	545	91
37	**Avocado**, *average*	4	14	Tr	16	3
38	*average, weighed with skin and stone*	3	10	Tr	11	2
39	Fuerte	4	14	Tr	16	3
40	*weighed with skin and stone*	3	10	Tr	11	2
41	Hass	4	14	Tr	16	3
42	*weighed with skin and stone*	3	10	Tr	11	2
43	**Babaco**	10	160	20	175	29
48	**Blackberries**, *raw*	4	78	0	80	13
49	stewed with sugar	3	61	0	62	10
50	stewed without sugar	3	66	0	68	11

		α- carotene	β- carotene	crypto- xanthins	Carotene equiv	Retinol equiv
		Carotenoid fractions				
51	**Blackberry and apple**, *stewed with sugar*	1	(39)	0	(40)	(7)
52	*stewed without sugar*	1	(43)	0	(44)	(7)
59	**Carambola**	24	25	0	37	6
60	**Cashew fruit**	5	55	80	115	19
61	**Cherries**, *raw*	4	23	0	25	4
62	*raw, weighed with stones*	3	19	0	21	3
67	*canned in syrup*	4	15	0	17	3
70	**Cherry pie filling**	2	17	0	18	3
71	**Clementines**	5	73	0	75	13
72	*weighed with peel and pips*	4	55	0	57	9
99	**Fruit salad**, *homemade*	1	19	0	20	3
101	**Gooseberries**, cooking, *stewed with sugar*	3	40	0	41	7
102	*stewed without sugar*	3	42	0	43	7
105	**Grapefruit**, *raw*	9	12	0	17	3
106	*raw, weighed with peel and pips*	6	8	0	11	2
118	**Guava**, *raw*	0	380	110	435	73
119	*raw, weighed with skin and pips*	0	340	100	390	65
123	**Kiwi fruit**	7	34	0	37	6
124	*weighed with skin*	6	29	0	32	5
125	**Kumquats**, *raw*	155	Tr	195	175	30
131	**Limes**, *peeled*	2	11	0	12	2
132	*peeled, weighed with peel and pips*	1	8	0	9	1
140	**Loquats**, *raw*	170	380	100	515	86
141	*canned in syrup*	31	70	19	95	16
146	**Mandarin oranges**, canned in juice	7	92	0	95	16
147	*canned in syrup*	7	105	0	105	18
150	**Mangoes**, ripe, canned in syrup	0	1455	34	1470	245
156	**Melon**, Canteloupe-type	16	990	0	1000	165
157	*weighed whole*	9	585	0	590	98
158	*weighed with skin*	11	655	0	660	110
171	**Nectarines**	Tr	45	26	58	10
172	*weighed with stones*	Tr	40	23	52	9

		Carotenoid fractions				
		α-carotene	β-carotene	crypto-xanthins	Carotene equiv	Retinol equiv
175	**Oranges**	8	24	Tr	28	5
176	*weighed with peel and pips*	6	17	Tr	20	3
178	**Passion fruit**	410	360	370	750	125
179	*weighed with skin*	250	220	230	460	77
180	**Paw-paw**, *raw*	0	130	1365	810	135
181	*raw, weighed with skin and pips*	0	97	1025	585	98
183	**Peaches**, *raw*	Tr	45	26	58	10
184	*raw, weighed with stone*	Tr	41	23	53	9
185	*dried*	Tr	345	200	445	74
186	*dried, stewed with sugar*	Tr	135	78	175	29
187	*-, stewed without sugar*	Tr	140	82	180	30
190	**Pears**, *average, raw*	0	17	3	18	3
191	*average, raw, weighed with core*	0	15	3	16	3
192	*-, raw, peeled*	0	18	3	19	3
193	*-, raw, peeled, weighed with skin and core*	0	13	2	14	2
196	*dried*	0	86	10	91	15
203	William, *raw*	0	20	10	25	4
204	*raw, weighed with core*	0	18	9	23	4
226	**Pomegranate**	26	20	0	33	5
227	*weighed with skin*	17	13	0	21	4
228	**Pomelo**	13	17	0	23	4
229	*weighed with peel and pips*	8	10	0	14	2
230	**Prickly pears**	20	30	10	45	7
231	**Prunes**	31	140	0	155	26
232	*weighed with stones*	26	120	0	130	22
233	*stewed with sugar*	14	66	0	73	12
234	*stewed with sugar, weighed with stones*	13	61	0	67	11
235	*stewed without sugar*	15	71	0	78	13
236	*stewed without sugar, weighed with stones*	14	65	0	72	12
237	canned in juice	15	135	0	140	23
238	canned in syrup	(15)	(135)	0	(140)	(23)
239	ready-to-eat	27	125	0	140	23
240	*weighed with stones*	23	110	0	120	20

		Carotenoid fractions				
		α- carotene	β- carotene	crypto- xanthins	Carotene equiv	Retinol equiv
257	**Satsumas**	5	73	0	75	13
258	*weighed with peel*	3	52	0	53	9
259	**Sharon fruit**	12	245	1395	950	160
264	**Tamarillos**	10	580	675	920	155
266	**Tangerines**	6	94	0	97	16
267	*weighed with peel and pips*	4	69	0	71	12
Fruit Juices						
281	**Orange juice**, freshly squeezed	(2)	(5)	(21)	(17)	(3)
282	freshly squeezed, *weighed as whole fruit*	(1)	(2)	(10)	(8)	(1)
283	unsweetened	2	5	21	17	3
284	**Orange juice concentrate**, unsweetened	19	53	200	170	29
285	**Passion fruit juice**	430	385	395	800	135
288	**Pomegranate juice, fresh**	(26)	(20)	0	(33)	(5)
Nuts and seeds						
849	**Trail mix**	3	45	0	47	8

VITAMIN E FRACTIONS

The vitamin E activity of foods can be derived from a number of different tocopherols and tocotrienols. In fruit, the major contributor is α-tocopherol, but γ-tocopherol is also important in some nuts.

Where vitamin E is present and the amount of each tocopherol was known, the values are shown below. For these fruit and nuts, the total vitamin E activity is also shown as α-tocopherol equivalents, which is the sum of the α-tocopherol, 40% of the β-tocopherol, 10% of the γ-tocopherol and 1% of the δ-tocopherol (McClaughlin and Weihrauch, 1979[a]).

		Vitamin E fractions, mg per 100g				
		α-tocopherol	β-tocopherol	γ-tocopherol	δ-tocopherol	Vitamin E equiv
Fruit						
48	**Blackberries**, *raw*	2.05	0	2.90	2.75	2.37
49	*stewed with sugar*	1.60	0	2.30	2.22	1.85
50	*stewed without sugar*	1.75	0	2.48	2.44	2.03
105	**Grapefruit**, *raw*	(0.19)	0	(0.01)	0	(0.19)
106	*raw, weighed with peel and pips*	(0.13)	0	(0.01)	0	(0.13)
173	**Olives**, *in brine*	1.97	0.02	0.12	Tr	1.99
174	*in brine, weighed with stones*	1.58	0.02	0.10	Tr	1.59
213	**Plums**, *average, raw*	0.60	0	0.07	0	0.61
214	*average, raw, weighed with stones*	0.56	0	0.07	0	0.57
215	*-, stewed with sugar*	0.50	0	0.06	0	0.51
216	*-, stewed with sugar, weighed with stones*	0.47	0	0.06	0	0.48
217	*-, stewed without sugar*	0.49	0	0.06	0	0.50
218	*-, stewed without sugar, weighed with stones*	0.47	0	0.06	0	0.47
219	*canned in syrup*	0.25	0	0.03	0	0.25
220	*Victoria, raw*	0.60	0	0.07	0	0.61
221	*raw, weighed with stones*	0.56	0	0.07	0	0.57
222	*stewed without sugar*	0.52	0	0.06	0	0.53
223	*stewed without sugar, weighed with stones*	0.49	0	0.06	0	0.50
224	*yellow, raw*	(0.60)	0	(0.07)	0	(0.61)
225	*raw, weighed with stones*	(0.58)	0	(0.07)	0	(0.59)

[a]McClaughlin, P. J. and Weihrauch, J. L. (1979) Vitamin E content of foods. *J. Am. Diet. Assoc.* **75,** 647–665

		α-tocopherol	β-tocopherol	γ-tocopherol	δ-tocopherol	Vitamin E equiv
244	**Raspberries**, *raw*	0.30	0	1.50	2.70	0.48
245	*stewed with sugar*	0.26	0	1.30	2.50	0.42
246	*stewed without sugar*	0.29	0	1.40	2.50	0.46
247	*frozen*	0.30	0	1.50	2.70	0.48
248	canned in syrup	0.10	0	0.50	0.90	0.15

Fruit juices

		α-tocopherol	β-tocopherol	γ-tocopherol	δ-tocopherol	Vitamin E equiv
275	**Grapefruit juice**, unsweetened	0.19	0	0.01	0	0.19
276	**Grapefruit juice concentrate**, unsweetened	0.70	0	0.02	0	0.70
281	**Orange juice**, freshly squeezed	(0.17)	0	(0.01)	0	(0.17)
283	unsweetened	0.17	0	0.01	0	0.17
284	**Orange juice concentrate**, unsweetened	0.68	0	0.03	0	0.68
287	**Pineapple juice concentrate**, unsweetened	0.14	0	0.01	0	0.14

Nuts and seeds

		α-tocopherol	β-tocopherol	γ-tocopherol	δ-tocopherol	Vitamin E equiv
801	**Almonds**	23.77	0.26	0.81	0	23.96
802	*weighed with shells*	8.79	0.10	0.30	0	8.86
803	*toasted*	24.17	0.26	0.82	0	24.36
807	**Bombay mix**	3.86	Tr	8.54	0	4.71
808	**Brazil nuts**	5.72	0.15	13.87	0.17	7.18
809	*weighed with shells*	2.63	0.07	6.38	0.08	3.30
811	**Cashew nuts**, *plain*	0.29	0.15	4.97	0.39	0.85
812	*roasted and salted*	0.77	0.04	5.09	0.38	1.30
813	**Chestnuts**	0.50	0	7.00	0	1.20
814	*weighed with shells*	0.41	0	5.81	0	0.99
815	*dried*	0.94	0	13.19	0	2.26
816	**Coconut**, *fresh*	0.70	0	0.30	0	0.73
817	*creamed block*	1.34	0	0.57	0	1.40
818	*desiccated*	1.21	0	0.52	0	1.26
819	**Coconut cream**	0.67	0	0.29	0	0.70
821	**Hazelnuts**	24.2	0.80	4.33	0.22	24.98
822	*weighed with shells*	9.20	0.30	1.65	0.08	9.49
825	**Marzipan**, *retail*	6.13	0.07	0.21	0	6.18
829	**Peanut butter**, smooth	4.70	Tr	2.90	Tr	4.99
831	**Peanuts**, *plain*	9.21	0.23	7.91	0.37	10.09
832	*plain, weighed with shells*	6.35	0.16	5.46	0.25	6.97
833	*dry roasted*	0.70	0.18	3.30	0.53	1.11
834	*roasted and salted*	0.41	0.14	1.90	0.37	0.66

		α-tocopherol	β-tocopherol	γ-tocopherol	δ-tocopherol	Vitamin E equiv
837	**Pecan nuts**	1.45	1.31	23.56	0.66	4.34
838	*weighed with shells*	0.71	0.64	11.5	0.32	2.12
839	**Pine nuts**	12.47	Tr	11.77	Tr	13.65
840	**Pistachio nuts**, *roasted and salted*	1.43	Tr	27.15	0.01	4.16
841	*roasted and salted, weighed with shells*	0.79	Tr	14.93	Tr	2.28
844	**Sesame seeds**	0.25	0	22.81	0.29	2.53
845	**Sunflower seeds**	37.20	1.20	0.92	0.34	37.77
846	*toasted*	(38.54)	(1.24)	(0.95)	0.35	(39.11)
847	**Tahini paste**	0.25	0	23.17	0.29	2.57
849	**Trail mix**	4.25	0	2.87	0	4.53
850	**Walnuts**	1.35	0.09	24.46	2.29	3.85
851	*weighed with shells*	0.58	0.04	10.52	0.98	1.66

PHYTIC ACID

Phytic acid is shown as grams of phytic acid per 100g edible portion of the food except where stated. Figures are from analyses, together with calculations from phytic acid phosphorus values on the basis that 1g phytic acid phosphorus is equivalent to 3.55g phytic acid.

		g per 100g

Fruit

37	**Avocado**, *average*	0.01	83	**Dates**, *raw*		0.04
38	*average, weighed with skin*		85	*dried*		0.09
	and stone	0.01	86	*dried, weighed with stones*		0.08
39	Fuerte	0.01	92	**Figs**, *dried*		0.03
40	*weighed with skin and stone*	0.01	95	*ready-to-eat*		0.03
41	Hass	0.01	148	**Mangoes**, ripe, *raw*		0.03
42	*weighed with skin and stone*	0.01	149	*raw, weighed with skin and*		
				stone		0.02
			242	**Raisins**		0.01

Nuts and seeds

801	**Almonds**	1.07	831	**Peanuts**, *plain*	0.77
802	*weighed with shells*	0.40	832	*plain, weighed with shells*	0.53
803	*toasted*	1.09	833	*dry roasted*	0.68
804	**Barcelona nuts**	0.25	834	*roasted and salted*	0.75
805	*weighed with shells*	0.15	835	**Peanuts, raisins and chocolate**	
808	**Brazil nuts**	1.32		**chips**	0.29
809	*weighed with shells*	0.61	836	**Peanuts and raisins**	0.44
811	**Cashew nuts**, *plain*	(0.97)	837	**Pecan nuts**	0.61
812	*roasted and salted*	0.97	838	*weighed with shells*	0.30
813	**Chestnuts**	0.01	840	**Pistachio nuts**, *roasted and*	
814	*weighed with shells*	0.01		*salted*	0.34
815	*dried*	0.02	841	*roasted and salted, weighed*	
816	**Coconut**, *fresh*	0.08		*with shells*	0.19
817	*creamed block*	0.14	844	**Sesame seeds**	1.38
818	*desiccated*	0.14	845	**Sunflower seeds**	3.00
821	**Hazelnuts**	1.07	846	*toasted*	3.11
822	*weighed with shells*	0.41	848	**Tigernuts**	0.64
823	**Macadamia nuts**, salted	0.29	850	**Walnuts**	0.58
			851	*weighed with shells*	0.25

ORGANIC ACIDS IN FRUIT

Fruits contain more organic acids than other food groups. The most widely occurring and abundant are citric and malic acids. Many other acids may be present, usually in much smaller amounts, although tartaric acid is a major acid in grapes and oxalic acid is present in significant amounts in rhubarb.

The concentrations of organic acids can be extremely variable depending on factors such as growing and storage conditions, ripeness, amount of sunlight received and even time of day.

The figures below are for fresh fruit and fruit juices only and have been selected from literature sources the majority of which were not from the UK. They are intended therefore only as a guide to the organic acid content of fruits in the UK and should be treated as such. Where no values are given the amounts present are likely to be negligible.

	Organic acids, g per 100g edible portion			
	Malic acid	Citric acid	Tartaric acid	Oxalic acid
Apples	0.5	Tr		
Apricots	1.1	0.4		Tr
Babaco	0.5	0.2		
Bananas	0.3	0.2		
Blackberries	0.6	0.9		Tr
Blackcurrants	0.4	2.9		
Carambola	0.2	0.8		0.3
Cherries	0.8	Tr	Tr	Tr
Figs, dried	0.2	0.3		
Gooseberries	0.7	0.7		
Grapes	0.5	Tr	0.5	Tr
Grapefruit	0.1	1.2		
Guava	0.2	0.6		
Kiwi fruit	0.2	1.0		
Lemons	0.1	4.6		
Limes	0.6	4.3		
Loganberries	0.5	1.9		
Loquats	0.5	Tr		
Lychees	0.3	Tr		
Mangoes	0.1	0.4	0.1	Tr
Melon, watermelon	0.3	0.1		
Mulberries	0.2	0.6		
Nectarines	0.6	0.4		
Oranges	0.2	1.0		
Passion fruit	0.5	2.5		
Paw paw	0.1	0.1		
Peaches	0.4	0.3		
Pears	0.2	0.1		Tr

	Organic acids, g per 100g edible portion			
	Malic acid	Citric acid	Tartaric acid	Oxalic acid
Pineapple	0.2	0.8		Tr
Plums	1.1	Tr		Tr
Pomegranate	0.1	1.0		
Prickly pears	0.3	0.1		
Quinces	0.8	Tr		
Raspberries	0.4	1.6		Tr
Redcurrants	0.3	2.1		Tr
Rhubarb	1.5	0.1		0.8
Strawberries	0.1	1.1		Tr
Tamarillos	0.2	1.8		
Grape juice	0.1	Tr	0.4	
Lemon juice	0.2	4.2		
Orange juice	0.2	1.0		

LITERATURE SOURCES FOR ORGANIC ACID LEVELS

Cashel, K., English, R. and Lewis, J. (1989) *Composition of foods, Australia. Volume 1.* Department of Community Services and Health, Canberra

Chisholm, D. N. and Picha, D. H. (1986) Effect of storage temperature on sugar and organic acid contents of watermelon. *Hort. Sci.* **21**, 1031-1033

Godinho, O. E. S., De Souza, N. and Aleixo, L. M. (1988) Determination of tartaric acid and the sum of malic and citric acids in grape juices by potentiometric titration. *J. Assoc. Off. Anal. Chem.* **71**, 1028-1031

Hulme, A. C. (1970) *The biochemistry of fruits and their products. Volume 1.* Academic Press. London and New York

Hulme, A. C. (1971) *The biochemistry of fruits and their products. Volume 2.* Academic Press. London and New York

Libert, B. (1981) Rapid determination of oxalic acid by reversed-phase high-performance liquid chromatography. *J. Chrom.* **210**, 540-543

MacRae, E. A., Lallu, N., Searle, A. N. and Bowen, J. H. (1989) Changes in the softening and composition of kiwifruit (*Actinidia deliciosa*) affected by maturity at harvest and postharvest treatments. *J. Sci. Food Agric.* **49**, 413-430

Nagy, S. and Shaw, P. E. (1980) *Tropical and subtropical fruits. Composition, properties and uses.* Avi Publishing Inc., Connecticut

Nicolas, J., Buret, M., Duprat, F., Nicolas, M., Rothan. C. and Moras, P. (1986) Effects of different conditions of cold storage upon physiochemical changes of kiwi fruit. *Acta Hort.* **194**, 261-272

Ramadan, A. A. S. and Domah, M. B. (1986) Non-volatile organic acids of lemon juice and strawberries during stages of ripening. *Die Nahrung* **30**, 659-662

Reyes, F. G. R., Wrolstad, R. E. and Cornwell, J. (1982) Comparison of enzymic gas-liquid chromatographic, and high performance liquid chromatographic methods for determining sugars and organic acids in strawberries at three stages of maturity. *J. Assoc. Off. Anal. Chem.* **65**, 126-131

Ruhl, E. H. (1989) Effect of potassium and nitrogen supply on the distribution of minerals and organic acids and the composition of grape juice of sultana vines. *Aust. J. Exper. Agric.* **29**, 133-137

Shaw, P. E. and Wilson, C. W. (1981) Determination of organic acids and sugars in loquat (*Eriobotrya japonica* Lindl.) by high-pressure liquid chromatography. *J. Sci. Food Agric.* **32**, 1242-1246

Souci-Fachman-Kraut (1987) *Food composition and nutrition tables 1986/87, 3rd revised and completed edition.* Wissenschaftliche Verlagsgesellschaft mbH, Stuttgart

Teles, F. F. F., Stull, J. W., Brown, W. H. and Whiting, F. M. (1984) Amino and organic acids of the prickly pear cactus (*Opuntia ficus indica* L.). *J. Sci. Food Agric.* **35**, 421-425

Wilson, C. W., Shaw, P. E. and Knight, R. J. (1982) Analysis of oxalic acid in carambola (*Averrhoa carambola* L.) and spinach by high-performance liquid chromatography. *J. Agric. Food. Chem.* **30**, 1106-1108

PROPORTIONS OF FRUIT IN CANS BEFORE DRAINING

In the main tables the values for canned fruit are for the fruit together with the syrup or juice in which it was canned. The figures below show the proportion of fruit present after the syrup or juice has been drained off. Where possible, for canned fruits with stones, values have been given for the stones present as well as for the edible portion only. In these cases the food number has been repeated in parentheses.

These values were determined experimentally by draining into a sieve. Ranges are included where available. In most cases the sample was that included in the main tables and was purchased from local outlets and supermarkets. Larger catering size cans may contain a different proportion of fruit to syrup or juice.

		Drained proportion	Range
34	**Apricots**, canned in syrup	0.64	(0.57 - 0.71)
35	canned in juice	0.64	(0.50 - 0.82)
58	**Boysenberries**, canned in syrup	0.53	
56	**Blackcurrants**, canned in juice	0.54	
67	**Cherries**, canned in syrup, no stones	0.47	(0.38 - 0.55)
(67)	canned in syrup, plus stones	0.61	
96	**Fruit cocktail**, canned in juice	0.65	(0.55 - 0.72)
97	canned in syrup	0.66	
104	**Gooseberries**, canned in syrup	0.53	
107	**Grapefruit**, canned in juice	0.52	(0.30 - 0.66)
108	canned in syrup	0.52	
120	**Guava**, canned in syrup	0.62	
136	**Loganberries**, canned in juice	0.49	
144	**Lychees**, canned in syrup	0.50	(0.45 - 0.63)
146	**Mandarin oranges**, canned in juice	0.56	(0.48 - 0.61)
147	canned in syrup	0.56	(0.52 - 0.67)
150	**Mangoes**, canned in syrup	0.61	
182	**Paw paw**, canned in juice	0.59	

		Drained proportion	Range
188	**Peaches**, canned in juice	0.68	(0.54 - 0.76)
189	canned in syrup	0.62	(0.58 - 0.68)
197	**Pears**, canned in juice	0.60	(0.54 - 0.63)
198	canned in syrup	0.61	(0.51 - 0.76)
211	**Pineapple**, canned in juice	0.54	(0.37 - 0.62)
212	canned in syrup	0.56	(0.48 - 0.63)
219	**Plums**, canned in syrup, no stones	0.45	(0.37 - 0.62)
237	**Prunes**, canned in juice, no stones	0.46	(0.42 - 0.49)
(237)	canned in juice, plus stones	0.51	
238	**Prunes**, canned in syrup, no stones	0.64	
(238)	canned in syrup, plus stones	0.71	
248	**Raspberries**, canned in syrup	0.52	
255	**Rhubarb**, canned in syrup	0.56	
262	**Strawberries**, canned in syrup	0.38	(0.33 - 0.46)

ALTERNATIVE AND TAXONOMIC NAMES

- Foods are listed below in the same order as in the main tables.
- The alternative names listed in the left-hand column below are those that were most frequently encountered during data collection and are included to help in identifying foods. It is important to recognise that in some cases such names may be used for more than one food and that all such usages may not appear in this list.
- To see if a name is listed, the food index should be consulted first. If the term is included as an alternative name, a cross reference entry (e.g.: Kula   see **Bananas**) indicates the food name to which it refers. This allows all alternatives to be listed together.
- Taxonomic names listed in the right-hand column refer as specifically as possible to the sources of data used. Where two or more taxonomic names are listed, the data are representative of a mixture of these varieties.
- The abbreviation 'var' is used to indicate the specific variety or unspecified variety(ies); 'sp' and 'spp' are used to indicate that one or more than one species of the specified Genus is included.

Alternative names	Food names	Taxonomic names
Fruit		
Indian gooseberry	**Amla**	*Emblica officinalis*
Tarel	**Apples**	*Malus pumila*
	Apricots	*Prunus armeniaca*
Alligator pear Butter pear Zaboca	**Avocado**	*Persea americana*
	Babaco	*Carica pentagona*
Kula	**Bananas**	*Musa* spp
Blueberries Huckleberries Whortleberries	**Bilberries**	*Vaccinium myrtillus*
	Blackberries	*Rubus ulmifolius*
	Blackcurrants	*Ribes nigrum*

Alternative names	Food names	Taxonomic names
	Boysenberries	*Rubus ursinus* var *loganobaccus*
Star apple Star fruit	**Carambola**	*Averrhoa carambola*
Christmas apple	**Cashew fruit**	*Anacardium occidentale*
	Cherries	*Prunus avium*
Acerolas Barbados cherries	**Cherries**, West Indian	*Malpighia punicifolia*
	Clementines	*Citrus reticulata* var Clementine
	Cranberries	*Vaccinium oxycoccus* *Vaccinium macrocarpon*
	Currants	*Vitis vinifera*
Corossol Netted custard apple	**Custard apple/Bullock's heart**	*Annona reticulata*
Cashiment Scaly custard apple Seethaphal Sweetsop	**Custard apple/Sugar apple**	*Annona squamosa*
Khezov	**Damsons**	*Prunus domestica* subsp *insititia*
	Dates	*Phoenix dactylifera*
Civet nut	**Durian**	*Durio zibethinus*
	Elderberries	*Sambucus* spp
Brazilian guava Pineapple guava	**Feijoa**	*Feijoa sellowiana*
Gullars	**Figs**	*Ficus carica*
	Gooseberries	*Ribes grossularia*
	Grapefruit	*Citrus paradisi*

Alternative names	Food names	Taxonomic names
	Grapes	*Vitis vinifera*
	Greengages	*Prunus domestica* subsp *italica*
Barbadines Yellow passion fruit	**Grenadillas**	*Passiflora edulis* f *flavicarpa*
	Guava	*Psidium guajava*
Waterapple	**Jambu fruit**	*Syzygium cumini* *Syzygium aqueum*
Chinese date Indian plum Zizyphus	**Jujube**	*Ziziphus jujuba*
Chinese gooseberry	**Kiwi fruit**	*Actinidia chinensis*
Cumquats	**Kumquats**	*Fortunella margarita* *Fortunella japonica*
	Lemons	*Citrus limon*
	Limes	*Citrus aurantiifolia*
	Loganberries	*Rubus loganobaccus*
Dragon eyes	**Longans**	*Nephelium longana* syn *Dimocarpus longan*
	Longans, canned	syn *Euphoria longan*
Japanese medlars Japanese plums	**Loquats**	*Eriobotrya japonica*
Chinese cherries Lichees Lichis Litchees Litchis	**Lychees**	*Litchi chinensis*
Big apricot Mamey	**Mammie apple**	*Mammea americana*

Alternative names	Food names	Taxonomic names
	Mandarin oranges	*Citrus reticulata*
	Mangoes	*Mangifera indica*
Mangistan	**Mangosteen**	*Garcinia mangostana*
	Medlars	*Mespilus germanica*
	Melon, Canteloupe-type	*Cucumis melo* var *cantaloupensis*
	Melon, Galia	*Cucumis melo* var *reticulata*
	Melon, Honeydew	*Cucumis melo* var *indorus*
	Melon, watermelon	*Citrullus lanatus*
	Mulberries	*Morus nigra*
	Nectarines	*Prunus persica* var *nectarina*
	Olives	*Olea europaea*
	Oranges	*Citrus sinensis*
	Ortaniques	*Citrus sinensis* x *Citrus reticulata*
Purple grenadilla	**Passion fruit**	*Passiflora edulis* f *edulis*
Papai Papaya	**Paw-paw**	*Carica papaya*
	Peaches	*Prunus persica*
	Pears	*Pyrus communis*
Bartlett pears	**Pears**, William	*Pyrus communis*
Asian pears	**Pears**, Nashi	*Pyrus pyrifolia*
Barbados gooseberry Cape gooseberry Chinese lantern	**Phyalis**	*Physalis peruviana*

Alternative names	Food names	Taxonomic names
	Pineapple	*Ananas comosus*
	Plums	*Prunus domestica* subsp *domestica*
Anar Granada	**Pomegranate**	*Punica granatum*
Pummelo Shaddock	**Pomelo**	*Citrus maxima* *Citrus grandis*
Cactus fruit Indian figs Tunas	**Prickly pears**	*Opuntia* spp
	Prunes	*Prunus domestica*
Japonicas	**Quinces**	*Cydonia vulgaris*
Kismis	**Raisins**	*Vitis vinifera*
Hairy lychee Rhambustan	**Rambutan**	*Nephelium lappaceum*
	Raspberries	*Rubus idaeus*
	Redcurrants	*Ribes rubrum*
	Rhubarb	*Rheum rhaponticum*
Chico Chiku Naseberry Noiseberry Sapota	**Sapodilla**	*Manilkara achras* syn *Achras sapota*
	Satsumas	*Citrus reticulata*
Chinese date plum Kaki Persimmon	**Sharon fruit**	*Diospyros kaki* syn *Embryopteris kaki*
	Strawberries	*Fragaria* sp

Alternative names	Food names	Taxonomic names
	Sultanas	*Vitis vinifera*
Tree tomatoes	**Tamarillos**	*Cyphomandra betacea*
	Tamarind	*Tamarindus indica*
	Tangerines	*Citrus reticulata*
	Whitecurrants	*Ribes sativum*

Nuts and seeds

Alternative names	Food names	Taxonomic names
Badam	**Almonds**	*Prunus amygdalus*
	Barcelona nuts	*Corylus maxima barcelonensis*
Areca nuts Betel-nut palms	**Betel nuts**	*Areca catechu*
	Brazil nuts	*Bertholletia excelsa*
Breadnuttree seeds	**Breadnut seeds**	*Brosimum alicastrum*
Kaju	**Cashew nuts**	*Anacardium occidentale*
	Chestnuts	*Castanea vulgaris*
Cocoanut Pharaoh's nut Nareal	**Coconut**	*Cocos nucifera*
	Hazelnuts	*Corylus avellana* *Corylus maxima*
Queensland nuts	**Macadamia nuts**	*Macadamia integrifolia* *Macadamia tetraphylla*
	Melon seeds	*Citrullus lanatus* *Cucumis melo*
Groundnuts Monkey nuts	**Peanuts**	*Arachis hypogaea*

126

Alternative names	Food names	Taxonomic names
Hickory nuts	**Pecan nuts**	*Carya illinoensis*
Indian nuts Pignolias Pine kernels	**Pine nuts**	*Pinus pinea*
Pista	**Pistachio nuts**	*Pistacia vera*
	Pumpkin seeds	*Cucurbita* spp
	Quinoa	*Chenopodium quinoa*
Benniseeds Gingelly Til	**Sesame seeds**	*Sesamum indicum*
	Sunflower seeds	*Helianthus annuus*
Chuffa Earth almonds	**Tigernuts**	*Cyperus esculentus*
Akhrots Madeira nuts	**Walnuts**	*Juglans regia*

REFERENCES TO TABLES

1   Caribbean Food and Nutrition Institute (1974) *Food composition tables for use in the English-speaking Caribbean.* Unwin Brothers, Woking

2   Cashel, K., English, R. and Lewis, J. (1989) *Composition of foods, Australia. Volume 1.* Department of Community Services and Health, Canberra

3   Gebhardt, S. E., Cutrufelli, R. and Matthews, R. H. (1982) *Composition of foods: fruits and fruit juices, raw, processed and prepared,* Agriculture Handbook No. 8-9, US Department of Agriculture, Washington DC

4   Gopalan, C., Rama Sastri, B. V. and Balasubramanian, S. C. (1980) *Nutritive value of Indian foods,* National Institute of Nutrition, Indian Council of Medical Research, Hyderabad

5   McCarthy, M. A. and Matthews, R. H. (1984) *Composition of foods: nut and seed products, raw, processed and prepared,* Agriculture Handbook No. 8-12, US Department of Agriculture, Washington DC

6   Polacchi, W., McHargue, J. S. and Perloff, B. P. (1982) *Food composition tables for the near east.,* Food and Agriculture Organization of the United Nations, Rome

7   Souci-Fachman-Kraut (1987) *Food composition and nutrition tables 1986/87, 3rd revised and completed edition.* Wissenschaftliche Verlagsgesellschaft mbH, Stuttgart

8   Visser, F. R. and Burrows, J. K. (1983) *Composition of New Zealand foods. 1. Characteristic fruits and vegetables.* DSIR Bulletin 235, New Zealand Department of Scientific and Industrial Research, Wellington

9   Wiles, S. J., Nettleton, P. A., Black, A. E. and Paul, A. A. (1980) The nutrient composition of some cooked dishes eaten in Britain: A supplementary food composition table. *J. Hum. Nutr.* **34**, 189-223

10   Wu Leung, W. T., Busson, F. and Jardin, C. (1968) *Food composition table for use in Africa,* Food and Agriculture Organization and US Department of Health, Education and Welfare, Bethesda

11   Wu Leung, W. T., Butrum, R. R., Chang, F. H., Narayana Rao, M. and Polacchi, W. (1972) *Food composition table for use in East Asia,* Food and Agriculture Organization and US Department of Health, Education and Welfare, Bethesda

12   Wu Leung, W. T. and Flores, M. (1961) *Food composition table for use in Latin America,* Institute of Nutrition of Central America and Panama, Guatemala City and Interdepartmental Committee on Nutrition for National Defense, National Institutes of Health, Bethesda

FOOD INDEX

- Foods are indexed by their food number and **not** by page number.

- Cross references in this index (e.g.: Kula   see **Bananas**) give access to the individual foods items through this index and to alternative names given in the Alternative and Taxonomic Names list on pages 121-127.